Carolina's Golden Fields

This book examines the environmental and technological complexity of South Carolina inland rice plantations from their inception at the turn of the seventeenth century to the brink of their institutional collapse at the eve of the Civil War. Inland rice cultivation provided a foundation for the South Carolina colonial plantation complex and enabled planters' participation in the Atlantic economy, dependence on enslaved labor, and dramatic alteration of the natural landscape. Moreover, the growing population of enslaved Africans led to a diversely acculturated landscape unique to the Southeastern Coastal Plain. Despite this significance, Lowcountry inland rice cultivation has had an elusive history. Unlike many historical interpretations that categorize inland rice cultivation in a universal and simplistic manner, this study explains how agricultural systems varied among plantations. By focusing on planters' and slaves' alteration of the inland topography, this book emphasizes how agricultural methods met the demands of the local environment.

HAYDEN R. SMITH is Adjunct Professor of History at the College of Charleston, South Carolina.

Cambridge Studies on the American South

Series Editors

Mark M. Smith, *University of South Carolina, Columbia*
Peter Coclanis, *University of North Carolina at Chapel Hill*

Editor Emeritus

David Moltke-Hansen

Interdisciplinary in its scope and intent, this series builds upon and extends Cambridge University Press's longstanding commitment to studies on the American South. The series offers the best new work on the South's distinctive institutional, social, economic, and cultural history and also features works in a national, comparative, and transnational perspective.

Titles in the Series

Hayden R. Smith, *Carolina's Golden Fields: Inland Rice Cultivation in the South Carolina Lowcountry, 1670–1860*
Wilson Jeremiah Moses, *Thomas Jefferson: A Modern Prometheus*
Joan E. Cashin, *War Stuff: The Struggle for Human and Environmental Resources in the American Civil War*
David Stefan Doddington, *Contesting Slave Masculinity in the American South*
Lawrence T. McDonnell, *Performing Disunion: The Coming of the Civil War in Charleston, South Carolina*
Enrico Dal Lago, *Civil War and Agrarian Unrest: The Confederate South and Southern Italy*
Daniel J. Vivian, *A New Plantation World: Sporting Estates in the South Carolina Lowcountry, 1900–1940*
Eugene D. Genovese, ed. Douglas Ambrose, *The Sweetness of Life: Southern Planters at Home*
Donald J. Mathews, *At the Altar of Lynching: Burning Sam Hose in the American South*
Keri Leigh Merritt, *Masterless Men: Poor Whites and Slavery in the Antebellum South*
Sarah Gardner, *Reviewing the South: The Literary Marketplace and the Southern Renaissance, 1920–1941*
Katherine Rye Jewell, *Dollars for Dixie: Business and the Transformation of Conservatism in the Twentieth Century*
Thomas Okie, *The Georgia Peach: Culture, Agriculture, and Environment in the American South*
Karlos K. Hill, *Beyond the Rope: The Impact of Lynching on Black Culture and Memory*
William A. Link and James J. Broomall, eds., *Rethinking American Emancipation: Legacies of Slavery and the Quest for Black Freedom*

James Van Horn Melton, *Religion, Community, and Slavery on the Colonial Southern Frontier*
Damian Alan Pargas, *Slavery and Forced Migration in the Antebellum South*
Craig Friend and Lorri Glover, eds., *Death and the American South*
Barton A. Myers, *Rebels against the Confederacy: North Carolina's Unionists*
Louis A. Ferleger and John D. Metz, *Cultivating Success in the South: Farm Households in Postbellum Georgia*
Luke E. Harlow, *Religion, Race, and the Making of Confederate Kentucky, 1830–1880*
Susanna Michele Lee, *Claiming the Union: Citizenship in the Post–Civil War South*
Ari Helo, *Thomas Jefferson's Ethics and the Politics of Human Progress: The Morality of a Slaveholder*
Kathleen Hilliard, *Masters, Slaves, and Exchange: Power's Purchase in the Old South*
Scott P. Marler, *The Merchants' Capital: New Orleans and the Political Economy of the Nineteenth-Century South*
Ras Michael Brown, *African-Atlantic Cultures and the South Carolina Lowcountry*
Johanna Nicol Shields, *Freedom in a Slave Society: Stories from the Antebellum South*
Brian Steele, *Thomas Jefferson and American Nationhood*
Christopher Michael Curtis, *Jefferson's Freeholders and the Politics of Ownership in the Old Dominion*
Jonathan Daniel Wells, *Women Writers and Journalists in the Nineteenth-Century South*
Peter McCandless, *Slavery, Disease, and Suffering in the Southern Lowcountry*
Robert E. Bonner, *Mastering America: Southern Slaveholders and the Crisis of American Nationhood*

Carolina's Golden Fields

Inland Rice Cultivation in the South Carolina Lowcountry, 1670–1860

HAYDEN R. SMITH
College of Charleston

CAMBRIDGE
UNIVERSITY PRESS

CAMBRIDGE
UNIVERSITY PRESS

University Printing House, Cambridge CB2 8BS, United Kingdom

One Liberty Plaza, 20th Floor, New York, NY 10006, USA

477 Williamstown Road, Port Melbourne, VIC 3207, Australia

314–321, 3rd Floor, Plot 3, Splendor Forum, Jasola District Centre, New Delhi – 110025, India

79 Anson Road, #06–04/06, Singapore 079906

Cambridge University Press is part of the University of Cambridge.

It furthers the University's mission by disseminating knowledge in the pursuit of education, learning, and research at the highest international levels of excellence.

www.cambridge.org
Information on this title: www.cambridge.org/9781108423403
DOI: 10.1017/9781108526005

First published 2020

Printed in the United Kingdom by TJ International Ltd. Padstow Cornwall

A catalogue record for this publication is available from the British Library.

Library of Congress Cataloging-in-Publication Data
NAMES: Smith, Hayden (Hayden R.), author.
TITLE: Carolina's golden fields : inland rice cultivation in the South Carolina Lowcountry, 1670-1860 / Hayden R. Smith.
DESCRIPTION: Cambridge, United Kingdom ; New York, NY : Cambridge University Press, 2020. | Series: Cambridge studies on the American South | Includes bibliographical references and index.
IDENTIFIERS: LCCN 2019012482 | ISBN 9781108423403 (hardback : alk. paper) | ISBN 9781108437523 (pbk. : alk. paper)
SUBJECTS: LCSH: Rice farming–South Carolina–History. | Rice farming–Environmental aspects–South Carolina. | Rice farming–Economic aspects–South Carolina. | Slave labor–South Carolina–History. | Plantations–South Carolina–History. | South Carolina–Economic conditions. | South Carolina–Environmental conditions.
CLASSIFICATION: LCC SB191.R5 S5996 2020 | DDC 338.4/766472509757–dc23
LC record available at https://lccn.loc.gov/2019012482

ISBN 978-1-108-42340-3 Hardback

For Katherine, Hayden Ros, Fulton, and Josephine

Contents

Figures

Table

Acknowledgments

Inspired by the often-cited African proverb, I have learned that it takes a community to complete a book. Without the patience, time, and effort from a large group of people, I would not have been able to complete this project. The original manuscript was a reflection of my dissertation at the University of Georgia. I am especially grateful for Paul Sutter, who guided me to transform a rough and wayward paper into a polished piece. Paul has an amazing ability to look beyond the information at hand and push his students and colleagues to form a relevant narrative that includes a broad historiography beyond environmental history. Allan Kulikoff also provided helpful and thoughtful analysis in this process. He asks tough questions and expects his students to understand the larger issues at stake. When I came to Athens, a young professor had just received an assignment in the history department. Shane Hamilton became an important part of my academic career, inspiring me to look at the important relationships between technology, culture, and food. Ben Ehlers has also provided a thoughtful role in guiding me to understand the relevance of plantation systems to the larger Atlantic world. Bea Burton, Christina Davis, Jim Gigantino, Daleah Goodwin, Darren Grem, Chase Hagood, Josh Haynes, Chris Huff, Chris Lawton, Robby Lucket, Jennifer Malto, Chris Manganiello, Jason Mathorne, Keri Leigh Merritt, LaShonda Mims, Barton Myers, Kathi Nehls, Tom Okie, Tore Olsson, Lesley-Anne Reed, Zach Smith, Drew Swanson, Levi VanSant, and Burt Way have all provided serious academic conversation and helped me strengthen this book over the years.

I am especially grateful to Richard Porcher, who initially encouraged me to research the importance of inland rice cultivation. Without his wisdom, this project would have never seen the light of day. He has spent a lifetime surrounded by the relics of South Carolina rice culture and knows better than anyone the subtleties of this bygone era. Through his career incorporating biology with history, Richard has become a pioneer in environmental history while humbly focusing on new endeavors. I am grateful that he took me under his wing to get me started in this project and introduced me to the expansive number of people who shared his interest in rice and history. I am especially thankful to the kind individuals who allowed me to walk their property and take time to help me see the remnants of the inland rice culture. Thank you to Jessica Loring and Laurence Rasmussen at Old Combahee, Larry Faulkenberry and Tom Milliken at Goodwill, Ernie Wiggers and Eddie Mills at Nemours, Carter Hudgins and staff at Drayton Hall. People who also shared a tremendous amount of time and knowledge are Jim and Esther Ferguson, Travis Folk, Linda Lear, and Cary Mock.

Part of the excitement of conducting research was meeting the wonderful and helpful staff of public and private institutions, both in the field and in the archives. I am especially grateful to Robert Morgan at Francis Marion National Forest. He opened up the US Forest Service (USFS) archives to me, showing all the overlooked inland rice features that still existed to this day. Because of Bob and his staff's help, most of my focus took place on USFS land. Carl Trettin and his staff at the Santee Experimental Forest also receive much appreciated salutations. Carl and Brooke James patiently put technical scientific information in layman's terms and helped explain the broader understanding of hydrology as it effected inland rice planting. I am also thankful to Mark Purcell and staff at the ACE Basin Nation Wildlife Refuge and Richard Kanaski at the US Forest and Wildlife Service. William Doar, III and the staff at the South Carolina Geological Survey patiently answered all queries pertaining the state's geology.

Invaluable assistance in the archives is attributed to the staff at the South Carolina Historical Society, especially to Jane Aldrich, Karen Stokes, Mary Jo Fairchild, and Neil Polhemus. Nic Butler and his helpful staff in the South Carolina Room at the Charleston County Public Library helped me reference much of their collections. The staff of the South Caroliniana Library and the South Carolina Department of Archives and History provided guidance in tackling the institution's overwhelming collections. Harriott Cheves Leland at the South Carolina Huguenot Society graciously provided transcribed information of early French settlers. The staff at the Charleston Library Society provided much help accessing their collections. John White and Simon Lewis of the Carolina Lowcountry and Atlantic World Program at the College of Charleston provided much needed assistance and a platform to present some findings to the general public. Finally, Robert McIntyre at the Charleston County Register Mesne Conveyance went above and beyond in allowing me access to the office's invaluable plat collections.

Drafts of this project went through untold revisions to polish arguments and clarity. Once again, I am grateful to the committee members in providing this assistance. I am also thankful to Theodore Rosengarten, Peter Hoffer, and Lyn Smith for taking time and energy to help proofread many of these chapters. This work also was strengthened by the helpful comments from participants at the WHEATS conference and New Histories of Rice workshop. Of the latter, I am grateful to Edda Fields-Black, Peter Coclanis, Francesca Bray, and Dagmar Schafer for inviting me to that remarkable event. At Cambridge University Press, a special thank you to Mark M. Smith and Peter Coclanis, the coeditors of the series Studies on the American South. I am especially grateful to the CUP staff for the tremendous amount of time and energy they spent to see this project through: Deborah Gershenowitz (Senior Editor), Rachel Blaifeder (Editorial Assistant), Sindhujaa Ayyappan (Project Manager) and Thomas Haynes (Content Manager). Finally, Belle Mundy's remarkable roll as copy editor honed all the rough edges in this work. Thank you.

Groups of people have watched my academic career grow over the decades and I would have been a lesser person academically without their guidance. I am especially grateful to Ted and Dale Rosengarten for taking me under their wing and honing my skills as an aspiring historian. Martha Zierden and Ron Anthony of the Charleston Museum have also been invaluable guides for over twenty years. Students of the epic Willtown dig that have been supportive over the years are Andrew Agha and Elizabeth Garrett Ryan. They have all been like family as our friendship grew from digging in heat and around mosquitoes.

Support from family is how I was able to finish this project. Without their love and support, I would have not made it this far. I am grateful to Greg and Lyn Smith and Brooke and Tobias Denny for patiently encouraging me to fulfill these goals. My in-laws, Mike and Sue Presley, and the Presley family have been overly supportive of this process. Finally, the person who has supported me the most in this process and to whom I cannot do justice for the patience and love that she gave to me over the years is my amazing wife, Katherine. She has sacrificed a lot of time and emotion to see this project completed. Katherine and our remarkable children have given me more joy and love during difficult times, making this life a remarkable adventure.

1

Introduction

In Land of Cypress and Pine

The basis for this book began twenty years ago when I enrolled in the College of Charleston's summer archaeological field school. After spending the first half of the semester honing our technique by digging five-foot by five-foot units, identifying soil stratigraphy, and collecting artifacts at the Charleston Museum's Stono Plantation, the archaeologists reoriented us students to a new site. For the remainder of the field school we investigated Willtown Bluff on the Edisto River, an early eighteenth-century township surrounded by plantations. My interest in inland rice cultivation grew from our work at the James Stobo site, a 1710 plantation located on the edge of the Willtown township and one mile from the tidal river. For three archaeological seasons between 1997 and 1999, I participated in excavations of the Stobo Plantation house foundation located on a hardwood knoll surrounded by a sea of low-lying cypress wetlands. During this time, I had a unique opportunity to walk off the dry terra firma and explore miles of inland rice embankments sprawling to the east and to the south of the house site. Major embankments traverse the wetlands on magnetic north/south and east/west axes, intersected by smaller check banks and drainage canals as far as the eye can see under the dense cypress and hardwood canopy.[1]

I was in awe of the expanse of Stobo's former inland rice fields and the tremendous amount of earth that enslaved laborers had moved to cultivate the grain. I also began to realize how this particular site was not

[1] Martha Zierden, Suzanne Linder, and Ronald Anthony, *Willtown: An Archaeological and Historical Perspective*, The Charleston Museum Archaeological Contributions 27 (Columbia: South Carolina Department of Archives and History, 1999).

thoroughly accounted for in the South Carolina rice historiography. Historical interpretations of inland fields note that the fields were simple in design and small in nature. As a result, I was not prepared to make sense of a site of this magnitude. I was also surprised by the sophistication of the water control techniques that Stobo's slaves utilized. Older historiography did not take into account the extensive canal networks used to channel water to and from the crop. Rather, historians had assumed that fields simply filled with impounded water that was then released downstream. At Stobo Plantation, slaves used a reservoir – or reserve – of impounded water to irrigate adjoining rice fields, which were embanked to retain water throughout the late spring and summer growing cycle. While noting the difference between text and landscape, I interpreted the Stobo site as exceptional, a high-water mark of this particular cultivation method.[2]

My curiosity about inland rice cultivation increased when I had the opportunity in 2005 to walk the most recent Francis Marion National Forest land purchase. The Charleywood tract, an eighteenth-century inland rice plantation, borders a tributary of the Wando River, which is also an arm of the Charleston Harbor. The plantation's rice fields provided a stark contrast to the Stobo site. While Stobo's fields followed the course of an inland small-stream floodplain and were irrigated by an

[2] Earlier histories note that inland cultivation was the first economically successful rice plantation. These interpretations also describe a single method of inland rice cultivation. See: Lewis Cecil Gray, *History of Agriculture in the Southern United States to 1860*, 2 vols (Washington, D.C.: Carnegie Institution of Washington, 1933; reprint, Gloucester, MA: Peter Smith, 1958), I: 280–284; Duncan Clinch Heyward, *Seed from Madagascar* (Chapel Hill: University of North Carolina Press, 1937; reprint, Columbia: University of South Carolina, 1993), 11–16; Converse D. Clowse, *Economic Beginnings in Colonial South Carolina, 1670–1730* (Columbia: University of South Carolina, 1971), 122–133; James M. Clifton, ed, *Life and Labor on Argyle Island: Letters and Documents of a Savannah River Rice Plantation, 1833–1867* (Savannah, GA: Beehive Press, 1978), ix–xi; Sam B. Hilliard, "Antebellum Tidewater Rice Culture in South Carolina and Georgia," in *European Settlement and Development in North America: Essays on Geographical Change in Honour and Memory of Andrew Hill Clark*, ed. James R. Gibson (Toronto: University of Toronto Press, 1978), 98–100; Peter A. Coclanis, *The Shadow of a Dream: Economic Life and Death in the South Carolina Low Country, 1670–1920* (New York: Oxford University Press, 1989), 44–45, 61–63, 96–98. Small scale inland rice cultivation continued beyond the traditional Lowcountry rice zone during the post-bellum period, documented in Amelia Wallace Vernon, *African Americans at Mars Bluff, South Carolina* (Baton Rouge: Louisiana State University Press, 1993), 73–128, and Peter A. Coclanis and J.C. Marlow, "Inland Rice Production in the South Atlantic States: A Picture in Black and White (African Americans in Southern Agriculture: 1877–1945)," *Agricultural History* 72 (Spring 1998): 197–212.

adjoining reservoir, the Charleywood fields spread over an expansive tidal riverbank and resembled the great "hydraulic machines" that made rice planters so wealthy. After a century and a half of agricultural abandonment, Guerin Creek's daily ebb and flow of brackish water inundated the impounded fields and caused the landscape to revert back to a spartina cordgrass ecosystem. How could scholars associate Charleywood with inland cultivation when its rice fields lie beside a tidal river? With no freshwater reservoir in sight of the former rice fields, I learned that Charleywood cultivators had to transport water two miles through canals to reach its agricultural destination. The rice fields looked like the fields bordering any of the tidal rivers throughout the South Carolina Lowcountry. Embankments surrounded the fields in a uniform and geometric manner and canals divided the fields to draw water onto and off the plants. Yet, this was classed as inland rice production because of the planter's dependence on reservoirs. With the overall popularity of tidal cultivation, nineteenth-century observers originally defined inland rice not by the distance from the ocean, but by the distance from tidal rivers. While this definition describes a majority of inland plantations, as this book points out, it is not always the case. I would define this cultivation strategy primarily by its dependence on the downward flow of water – from the water source, through the rice fields, and out to a tributary or river. Through my observations at Stobo and Charleywood plantations, and among many other inland field systems during this period, I realized that the story of inland rice cultivation was much more complex and varied than historians had previously realized.

Reservoir-irrigated rice cultivation was the first successful type of plantation agriculture developed in South Carolina and it served as the foundation for the South Carolina colonial economy. But despite its significance, Lowcountry inland rice cultivation has had an elusive history. Unlike the visible tidal rice embankments still standing along South Carolina tidal rivers, remnant inland fields are harder to find and many presently lie in overgrown wooded watersheds. The lack of cultivation has transformed the once carefully managed fields into second- or third-growth forests and wetlands, some of which are protected as conservation lands today in ways that have obscured these past histories of human land use. The sparseness of primary records also has deterred historians from fully examining the impact that this early plantation complex had on Lowcountry history, as few plantation journals and ledger books survive from the colonial period that speak of inland rice culture. When tidal rice irrigation took hold in the mid-eighteenth century, most planters began focusing their slave labor and documentation on this new technology because of the efficiency in irrigation and higher yields. Yet, inland cultivation continued in the antebellum period, as evidence from nineteenth-century plats and journals make clear. Far from being a primitive early approach to rice growing, inland cultivation has a history that parallels and interweaves with that of tidal cultivation.

This book argues the importance of how planters both adapted to and altered their environment by planting rice in South Carolina's inland swamps during the colonial and antebellum periods. It shows how attention to the environment leads to a historical analysis of the Lowcountry cultivators and the land. I ask why people chose to pursue such a labor-intensive crop, and how aspiring planters viewed the blank Lowcountry canvas and implemented their technological understanding of the world in constructing irrigation systems and agricultural fields. Inland cultivation began as a simple process for growing rice by taking advantage of suitable sites. As demand for the crop and land values increased, planters needed larger harvests and so spent more energy expanding old inland fields and crafting new inland rice environments. The need to adapt to the diverse landscapes of the South Carolina Lower Coastal Plain prompted planters to make each tract unique in order to maximize available land for rice cultivation. Their enslaved cultivators worked within the limitations of this environment to manage water flow and lessen the impact of storms, flooding, and drought, but as time went by they also transformed these environments in increasingly sophisticated ways.

The second theme of this book emphasizes the broader significance of inland rice, as one cannot understand how Lowcountry rice culture (and the broader understanding of the plantation economy) developed without appreciating its origins in inland planting. This argument provides a significant story about slave labor systems in the Americas. When experimenting initially with rice cultivation, colonists used African slaves to plant seeds in a variety of microenvironments. Reacting to the opportunities of the global economy, inland planters used enslaved labor continually to clear more land and expand the crop's output, just as tobacco planters in the Chesapeake set slaves there to clear new land. This practice encouraged the ever-expanding slave trade in South Carolina and the diaspora of Africans through the New World. Ways of mobilizing labor by task instead of by gang also took shape on inland rice plantations. The task system that developed in the Lowcountry is found nowhere else in American history. The ecological foundations of inland rice plantations are the keys to understanding the emergence of a highly intricate labor and environmental management system in the dense South Carolina woodland watersheds.[3]

By the first decade of the eighteenth century, rice had become South Carolina's most successful commodity, partially due to the conducive southeastern geography. Freshwater wetlands ideal for this cultivation were nestled in tributaries and swamps several miles away from the South Carolina coast. Geographer Judith Carney describes these wetlands as "an array of microenvironments which include valley bottoms, low-lying depressions, and areas of moisture holding clay." The inland terrain challenged planters to recognize what features would successfully sustain rice. Once they identified these features, planters used enslaved Africans to revamp these available natural features for cash crop production. From hardwood depressions down to the cypress riverbanks, various ecosystems within the Lowcountry were modified for inland swamp plantations. The critical requirement for widespread cultivation was active water flow through these landforms.[4]

For the purposes of this book, the South Carolina Lowcountry begins at the coastline of the Lower Coastal Plain and extends approximately fifty miles inland. From a geological perspective, the Lowcountry boundary extends to the Surry Scarp. From a political perspective, the boundary was the inland survey line of the eighteenth-century Anglican parishes. Whether a coincidence or reflective of how geological formations had an effect on political boundaries, both lines of demarcation line up to approximately the same distance from the coastline. I identify the Lowcountry as a proper region similar to the South, so in this case I capitalize the term.[5]

[3] Philip D. Morgan, "Work and Culture: The Task System and the World of Lowcountry Blacks, 1700 to 1800," *William and Mary Quarterly*, 3rd ser., 39 (October 1982): 563–599; Peter A. Coclanis, "How the Lowcountry Was Taken to Task: Slave-Labor Organization in Coastal South Carolina and Georgia," in *Slavery, Secession, and Southern History*, eds. Robert Louis Paquette and Louis Ferleger (Charlottesville: University Press of Virginia, 2000), 59–78.

[4] Judith Carney, *Black Rice: The African Origins of Rice Cultivation in the Americas* (Cambridge, MA: Harvard University Press, 2001), 58; Hilliard, "Antebellum Tidewater Rice Culture," 97; S. Max Edelson, "Clearing Swamps, Harvesting Forests: Trees and the Making of a Plantation Landscape in the Colonial South Carolina Lowcountry," *Agricultural History* 81 (Summer 2007): 390–393.

[5] Charles F. Kovacik and John J. Winberry, *South Carolina: The Making of a Landscape* (Boulder: Westview Press, 1987; reprint, Columbia: University of South Carolina Press, 1989), 20–21; Donald J. Colquhoun, *Terrace Sediment Complexes in Central South Carolina* (Columbia: University of South Carolina, 1965), 21–23, 26.

Inland rice cultivation depended upon the simple principle that water flows from high ground to low ground. Water dispersed from rainstorms and springs flowed downhill, of course, while watersheds pulled this resource into creeks and streams. Inland planters found land in the Lowcountry that was level enough for rice cultivation, yet with a sufficient angle of two to three percent grade to allow drainage. Inland rice fields soon took shape throughout the South Carolina Coastal Plain, mostly in areas that were naturally suited to them. The physiographical coastal plain is generally downward sloping from the edge of the North American tectonic plate, the "fall line," to the Atlantic Ocean shoreline. The Lowcountry topography provided ideal situations for inland rice cultivation. As the Atlantic Ocean's shoreline alternately encroached and retreated during the Pleistocene epoch (approximately 2 million to 10,000 years ago), barrier island chains and corresponding tidal flats formed over the millennia to create terraces and scarps. Similar to modern barrier island systems, prehistoric terraces consisted of sand and shells, while the backside of these ridgelines consisted of clay loam from former tidal marshes and lagoons. Scarps serve as physical lines of demarcation – a transition – between the terraces, forming either from erosion of the receding coastline or during the depositional stage of former barrier islands.[6] Water's movement through these sedimentary deposits shaped the land, forming knolls, ridges, and troughs between four and forty feet in elevation, which became critical features to rice plantations and the people who lived on them. Islands of "high pine land" lying just a few feet within and around plantation swamps provided sites for buildings and grazing fields, while creeks flowing around these landforms provided the water sources and floodplains needed for cultivating rice. The early agricultural practices were of necessity diverse, as planters adapted their economic activities to the various microenvironments located on their property. Rather than altering their environments extensively, early inland rice planters used the environments that they found.[7]

[6] Donald J. Colquhoun, *Geomorphology of the Lower Coastal Plain of South Carolina* (Columbia: Division of Geology, 1969), 23, 6; David R. Soller and Hugh H. Mills, "Surficial Geology and Geomorphology," in *The Geology of the Carolinas: Carolina Geological Society Fiftieth Anniversary Volume*, eds. J.W. Horton, Jr. and V.A. Zullo (Knoxville: University of Tennessee Press, 1991), 290–291.

[7] Kovacik and Winberry, *South Carolina*, 20–21.

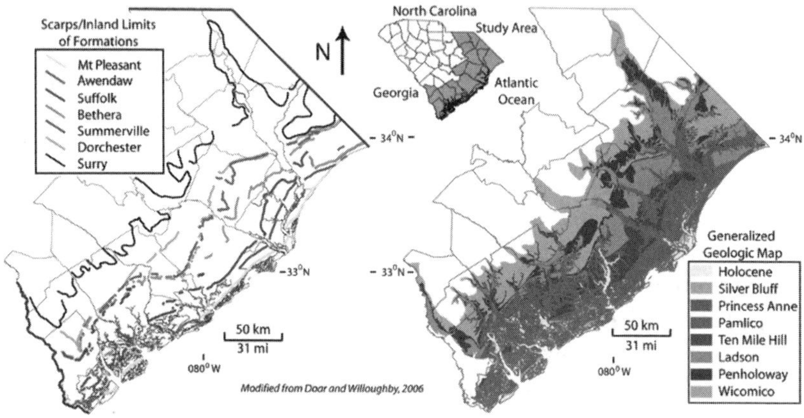

FIGURE 1.1 Generalized scarp and geology maps of the South Carolina Coastal Plain.

Image in William Richardson Doar, III, "The Geologic Implications of the Factors that Affected Relative Sea-Level Positions in South Carolina during the Pleistocene and the Associated Preserved High-stand Deposits" (Ph.D. diss., University of South Carolina, 2014), 58

FIGURE 1.2 The South Carolina Lowcountry represented in Anglican parishes, ca.1768. Dates indicate when parishes were founded.

Image from the South Carolina Department of Archives and History. http://archivesindex.sc.gov/guide/CountyRecords/parishes.htm (accessed April 5, 2019)

It is this struggle with the natural world that defined inland rice planters and dictated their agricultural decisions. Origins of these tensions appear in the Chapter 2, which discusses early rice cultivation strategies in South Carolina from the grain's approximate introduction in 1685 to the end of the proprietary period in 1729. During this time, colonists transformed the grain from one of several experimental commercial ventures into the central cash crop of early colonial South Carolina. Domestic and international demand for rice motivated colonists to seek out the best methods to grow and process ever-greater quantities of this non-native crop. Planter dependence on enslaved labor to clear land, create field infrastructure, and sow and harvest the crop, increased. In this chapter, I discuss the dynamic relationship of rice farming with topography and culture. European colonists began experimenting with rice cultivation alongside that of wheat and barley. At the same time, Africans knowledgeable about growing rice made it thrive in wetland areas and it provided the necessity of subsistence gardens. By the turn of the eighteenth century, these two cultural interpretations of rice farming merged to produce grain on small-stream floodplains. At the heart of this chapter is an analysis of how both free and enslaved people used various topographies to cultivate a particular grain and the lasting results that evolved from the early plantation landscape.

Chapter 3 discusses the dramatic transformation of inland rice cultivation between 1730 and the end of the American Revolution which coincided with the appearance of tidal irrigation. A combination of the reopening of the colonial land office and the relatively stable price of exported rice created a surge in land acquisitions that moved further into the South Carolina frontier. Spurred on by the land boom, planters moved rice cultivation from small-stream floodplains down to broad inland basins. Their shift in topographical focus required planters to construct more intricate canal and embankment systems to move larger volumes of water on and off the rice fields. In order to build elaborate infrastructures on these low-lying wetlands, planters had to invest in additional enslaved labor. Planters forced slaves to dig out tremendous amounts of earth to create channels for water and use the soil to build networks of dams and embankments. I argue the dramatic change in inland rice cultivation was modeled on planters' development of tidal irrigation along the Lowcountry rivers throughout the mid-eighteenth century. Both the evolving inland system and emerging tidal system required more extensive labor forces than before to create precisely leveled fields, massive embankments, and extensive canals. Creating a more extensive irrigation and drainage network called for a sophisticated understanding of hydrology and soils. With the intense development of rice fields in the Lowcountry basins, inland planters also encountered new problems. Malaria, declining soil fertility, pests, freshets and droughts all documented how the natural environment and the built environment could work at cross-purposes. Solutions to these conflicts were not in place when the American Revolutionary War put a temporary halt to rice output. The majority of planters abandoned rice cultivation during the Revolution and inland plantations were left in disrepair.

The latter chapters in this book detail the evolution of inland rice plantations from the end of the American Revolution until the Civil War. Chapter 4 highlights the collective effort of four rice plantations on the Wando River headwaters in Charleston County that enabled the owners to cultivate the crop up to the start of the Civil War. Current historiography discounts the role of inland rice cultivation in the antebellum period and focuses on the dominant tidal system. This chapter explains how inland cultivation maintained an important presence in the Lowcountry landscape. Indeed, contrary to inland rice's reputation as primitive, inland planters actively contributed to emerging trends in the scientific management of cash crops. To illustrate the complex role that inland rice plantations played in contrast to the predominant tidal system, this chapter provides a microanalysis of these four inland plantations – Charleywood, Fairlawn, Clayfield, and Wythewood – from 1783 to 1860. The owners of these tracts aggressively annexed surrounding plantations, intensified water management through canalization, and maintained a substantial enslaved labor population to carry out these tasks. Highlighting these four plantations, this chapter traces the evolution of inland rice culture and describes how it resembled and then diverged from tidal cultivation practices. Most inland planters, realizing the limitations of their soil fertility and of reliable impounded water, made adjustments in sowing techniques and flood schedules to increase irrigation efficiency. Having limited natural resources required inland planters to give additional attention to subtle changes in weather and environment. By 1840, these four plantations had been acquired by a new generation of planter entrepreneurs who sought to capitalize on their prior successes and diversify their rice holdings. However, these new planters had difficulty balancing the impounded water with successful cultivation as they watched their investments decline in rice output and property value before succumbing to total disruption from the Civil War.

In contrast to the fourth chapter's focus on the large antebellum inland rice planters, Chapter 5 explains how aspiring planters used inland rice plantations as a way of entering the planter aristocracy before the Civil War. Land values, as they relate to shifts in technology and agricultural output, paved the way for an emerging upper middle class to enter into the planter elite. During the antebellum period, most productive rice lands were beyond the means of professionals and merchants striving to enter into the upper echelons of society. When put up for sale, tidal rice plantations received a premium price, and the most desired lands stayed in families through inheritance or marriage. Land, and rice production, was a means to reflect one's status and define one's title in the rigid Lowcountry social hierarchy. Inland rice plantations, on the other hand, were more affordable and did become available to people aspiring to obtain rice planter status, although attempting this mode of social elevation came with monetary and emotional costs. Planters still had to populate their fields with a labor force, often in limited numbers, while the inland environment made difficult any attempts to plant the grain.

Chapter 6 documents inland cultivation strategies during the final two decades of the antebellum period. Using as a model the Biggin Basin, located at the headwaters of the Cooper River, this chapter discusses how a community of former inland rice planters revitalized the practice to supplement cotton production as a way to counter the fluctuating market. Revival of inland rice was a consequence of agricultural reform that took hold in select planter circles in the mid-nineteenth century. Lowcountry planters were part of this larger population having received the message through agricultural journals, scientific books, and agricultural societies. Promoters of agricultural reform called for a modern and scientific practice of agriculture to maintain soil fertility and crop output, halt westward migration, and curb the loss of status and political power by the South Atlantic states. Despite the lukewarm reception given to scientific agriculture in the Lowcountry, Biggin Basin planters began practicing conservation methods and expanding their operations. In this uncertain time, inland rice cultivation became a symbol of success and represented the very cash crop that brought wealth and status to this region a century earlier. Yet, for much of inland rice production on the eve of the Civil War, the realities of the market, labor requirements, and environmental limitations discouraged many from considering inland rice as a viable alternative to their more familiar cotton crop. By 1860, Biggin Basin planters abandoned any possibility of agricultural diversity by focusing on cotton production.

Ultimately, the story of inland rice cultivation in the South Carolina Lowcountry reflects the changing role of land use in relation to technology and culture. Inland rice cultivation took hold as the premier method of producing early colonial South Carolina's cash crop, and it evolved with further developments in land alteration and water management. However, when disease and agricultural disasters started to reveal the shortcomings of this built environment, and when the law turned against them, planters relegated inland rice cultivation to a secondary status. After the Revolution, those planters who endorsed its use defined inland rice cultivation in new ways. For some planters, inland rice was a legacy of agriculture initiated by their ancestors, but outdated for the times. Other planters practiced this older mode of rice cultivation out of desperation to achieve social or economic stability: for example, rice became a mode with which to improve their economic stability during fluxes in the long-staple cotton market. As shown in Chapter 7, to formerly enslaved peoples whose ancestors had worked the inland rice plantations for almost two centuries and also maintained small fields for themselves, this type of cultivation represented both oppression and subsistence. Inland rice cultivation represents a model of human interaction with the land. This book examines the agricultural practice to show how exchange and interconnection between people and their environment created both intentional and unforeseen effects.

2

Simple Reserves

Early Development of Inland Rice, 1670–1729

The origins of inland rice cultivation are as murky as the remnant fields today: overgrown in lore or overdeveloped into obscurity. From a physical perspective, one would have difficulty locating these early field systems: they exist as one travels out of Charleston. Following the rivers that formed the desirable peninsula to Native Americans and later colonists, one would have to look with precision to find an example that existed before the Stono Rebellion in 1739. Traces of early inland fields exist at Drayton Hall on the Ashley River, Caw Caw Country Park, Dungannon Plantation Heritage Preserve, throughout Francis Marion National Forest, and the fortunate private landholdings. However, for the most part, early examples that pay homage to the tremendous amount of enslaved labor and suffering used to move massive amounts of soil to eventually perfect irrigation canals, drainage ditches, impoundments and dams have disappeared, along with the primary sources that once meticulously recorded said events. Most of these earthworks have eroded into the landscape, overgrown with vegetation, or have succumbed to modern development.

This chapter discusses the rise of the South Carolina plantation complex and the evolution of rice cultivation from colonial settlement in 1670 to the collapse of the provincial government in 1729. During this sixty-year period, rice cultivation evolved from simple plots to intricate irrigated fields. While this chapter does not tackle the specifics of how rice was introduced in Carolina, it does emphasize how local ecological zones and global economic networks shaped the cultural identities and practices of Europeans and Africans in the New

World.[1] Attention to environmental conditions provides an important understanding of how people grew rice during this period. Colonists and slaves adapted their farming techniques to individual environments and by observing which land worked best to grow rice, colonists made rice the most successful cash crop in the Lowcountry. By understanding how rice cultivation changed during this initial period, a better understanding of inland rice production and its close relation to the environment will become more apparent.[2] This chapter discusses how European and African rice farmers initially cultivated rice in two separate ecosystems. By the end of the proprietary era, these two cultivation practices represented – to borrow Richard White's phrase – a "middle ground" on the small-stream floodplains.[3]

To understand how rice planting took shape in the early colony, this chapter integrates cultural perceptions of the landscape with the technological development that emerged during this period. Topography, defined by soil and water, provided a foundation for early colonial

[1] For examples of early histories on the origins of rice in Carolina, see Robert F.W. Allston, *Memoir of the Introduction and Planting of Rice in South Carolina* (Charleston: Miller & Brown, 1843); Alexander Samuel Salley, Jr., *The Introduction of Rice Culture in South Carolina*, Historical Commission of South Carolina, Bulletin, No. 6 (Columbia: State Company, 1919); Duncan Clinch Heyward, *Seed from Madagascar* (Chapel Hill: University of North Carolina Press, 1937; reprint, Columbia: University of South Carolina Press, 1993); Lewis Cecil Gray, *History of Agriculture in the Southern United States to 1860*, vol. 2 (Washington, D.C.: Carnegie Institution of Washington, 1933; reprint Gloucester, MA: Peter Smith, 1958).

[2] For recent scholarship documenting the origins of rice in Carolina, see Converse D. Clowse, *Economic Beginnings in Colonial South Carolina, 1670–1730* (Columbia: University of South Carolina Press, 1971); Peter H. Wood, *Black Majority: Negroes in Colonial South Carolina from 1670 through the Stono Rebellion* (New York: Alfred A. Knopf, 1974); James M. Clifton, "Introduction," in *Life and Labor on Argyle Island: Letters and Documents of a Savannah River Rice Plantation*, ed. James M. Clifton (Savannah, GA: Beehive Press, 1978), vii–xlvi; Daniel Littlefield, *Rice and Slaves: Ethnicity and the Slave Trade in Colonial South Carolina* (Baton Rouge: Louisiana State University Press, 1981); Peter A. Coclanis, *The Shadow of a Dream: Economic Life and Death in the South Carolina Low Country, 1670–1920* (New York: Oxford University Press, 1989); Philip D. Morgan, *Slave Counterpoint: Black Culture in the Eighteenth-Century Chesapeake and Lowcountry* (Chapel Hill: University of North Carolina Press, 1998); Judith A. Carney, *Black Rice: The African Origins of Rice Cultivation in the Americas* (Cambridge, MA: Harvard University Press, 2001); S. Max Edelson, *Plantation Enterprise in Colonial South Carolina* (Cambridge, MA: Harvard University Press, 2006); Judith A. Carney and Richard Nicholas Rosomoff, *In the Shadow of Slavery: Africa's Botanical Legacy in the Atlantic World* (Berkeley: University of California Press, 2009).

[3] The concept "middle ground" is found in Richard White, *The Middle Ground: Indians, Empires, and Republics in the Great Lakes Region, 1650–1825* (New York: Cambridge University Press, 1991).

settlement patterns and agricultural experimentation. Influenced by cultural perceptions and agricultural knowledge, people interpreted the new-world landscape in a variety of ways. This chapter discusses how evolving European-based perceptions of wetlands affected, and were affected by, the emerging rice culture. Although early examples of inland rice plantations are rare, the final section of this chapter considers four models of early reservoir-irrigated rice plantations in the Lowcountry and explains how each plantation varied subtly according to the microenvironment.

By the third quarter of the eighteenth century, rice production affected almost every aspect of colonial South Carolina's economy, politics, culture, and environment. Yet despite the grain's impact, rice began inconspicuously as one of a dozen experimental crops. While free colonists and enslaved laborers carved out cultivable land, Europe's demand for rice established South Carolina as one of the most economically successful colonies by the eve of the American Revolution. By 1774, the mean total wealth per inventoried estate (of free whites) in the Charleston district was £2,338 sterling. In comparison, Anne Arundel County, Maryland produced the second highest mean total of £660 sterling. Of the ten wealthiest men in British North America, nine were from the South Carolina Lowcountry. According to Peter Coclanis and Lacy Ford, "no area planted by Europeans in North America during the early modern period experienced such an impressive economic rise as South Carolina."[4] Rice had such a high value during the late colonial and antebellum periods that planters used it to pay for their children's education overseas and buy more slaves or land. By the beginning of the antebellum period, regionally grown rice had taken the name "Carolina Gold" from the description of its yellow husk and its value in worldwide markets. The desire for wealth and status motivated many families to migrate to this semitropical place, where they battled heat, disease, and hurricanes so they might try their luck with the risky but profitable endeavor.

Rice did not grow in the Lowcountry at the outset of colonization. The grain is a non-native species to the Americas, and colonists had to carefully tend to the plant in order for it to thrive. Colonists responded to the

[4] Peter A. Coclanis and Lacy K. Ford, "The South Carolina Economy Reconstructed and Reconsidered: Structure, Output, and Performance, 1670–1985," in *Developing Dixie: Modernization in a Traditional Society*, eds. Winfred B. Moore, Jr., Joseph F. Tripp, and Lyon G. Tyler, Jr. (New York: Greenwood Press, 1988), 94; Gray, *Agriculture in the Southern United States*, 752, 1030; Coclanis, *Shadow of a Dream*, 90–91; Alice Hanson Jones, *Wealth of a Nation to Be: The American Colonies on the Eve of the Revolution* (New York: Columbia University Press, 1980), 170–171, 377–379.

Lord Proprietors' call for agricultural prosperity by cultivating an array of desirable crops, yet the new inhabitants failed at many of these attempts because the Lowcountry climate could not support Mediterranean staples like olives and grapes, or new-world desirables such as cocoa. Agricultural experimentation took place on varied terrain, as diverse ecosystems existed within plantation boundaries. The proprietary tracts – no more than forty miles from the Atlantic coastline – included an assortment of geographical features, from dry upland ridges to wet low-lying troughs. From the outset of colonization, Lord Proprietor Anthony Ashley Cooper optimistically instructed colonists to plant "cotton seed, indigo seed, [and] ginger roots" in a variety of soils, for "our reason for this is that being unacquainted with ye nature of ye soyle [*sic*], we shall have conveniency of trying which sort of soile [*sic*] agrees best with ye severall [*sic*] things planted in them."[5]

With each wave of immigration, settlers fanned out from Charles Town following navigable waterways into the Carolina frontier. By 1690, colonists claimed land along the Ashley and Cooper rivers plus the navigable tributaries of the Stono River, Goose Creek, and Back River.[6] Under the headright system, the head of the household received 150 acres for every free person and male servant over sixteen years old plus an additional 100 acres for every male servant under sixteen years old and each woman servant regardless of her age. Although the Proprietors initially recruited colonists who were "seasoned" from living in the West Indies, the new arrivals had a difficult time producing commodities in the new soil. The seventeenth-century Carolina plantation economy faltered, however, because of limited agricultural knowledge conducive to the Lowcountry environment and too few workers to transform the landscape.[7] Natural disasters throughout the 1670s, with summer

[5] "Copy of Instruccons for Mr. West about our Plantacon," Langdon Cheves, ed., *The Shaftesbury Papers and Other Records Relating to Carolina* (South Carolina Historical Society, 1897; reprint, Charleston: Tempus Publishing, Inc., 2000), 126.

[6] Charles F. Kovacik and John J. Winberry, *South Carolina: The Making of a Landscape* (Columbia: University of South Carolina Press, 1989), 68–69; Meaghan N. Duff, "Creating a Plantation Province: Proprietary Land Policies and Early Settlement Patterns," in *Money, Trade, and Power: The Evolution of Colonial South Carolina's Plantation Society*, eds. Jack P. Greene, Rosemary Brana-Shute, and Randy J. Sparks (Columbia: University of South Carolina Press, 2001), 7–16; John S. Otto, *The Southern Frontiers, 1607–1860: The Agricultural Evolution of the Colonial and Antebellum South* (New York: Greenwood Press, 1989), 29.

[7] Linda M. Pett-Conklin, "Cadastral Surveying in Colonial South Carolina: A Historical Geography" (Ph.D. diss., Louisiana State University, 1986), 21–22; Meaghan N. Duff,

droughts and freezing winters, created a series of crop failures. From the outset, colonists faced environmental difficulties in growing crops. In the first year of colonization a late October freeze killed all of their crops "before they could come to perfection."[8] The next spring a prolonged drought killed all subsistence crops along with experiments in ginger and indigo. By the second year of colonization, colonists quickly learned that the Charleston climate was not consistent with Barbados, which many had used mainly as a referent. The "sharp and cold" winters, according to one colonist, killed "any thing of a Comodity [*sic*]," including Barbadian imports of sugar cane, cotton, and ginger.[9] As they came to understand the subtleties of soil and weather, early colonists had to make shifts in their cultivation strategies in response to environmental realities.

Despite the challenges that colonists faced with poor crop output during the first decade of colonization, they described the Carolina landscape with optimism. After the devastating 1670/1 winter, one colonist wrote of a "winter soe [*sic*] mild & temperate yet it may rather be termed a continuall [*sic*] spring." Although the author suffered through debilitating crop failures, he still believed Carolina was the "Land of Canaan, the habitation of the then elect & chosen people of god it is a Land flowing with milk & honey."[10] Seventeenth-century promotional tracts also pictured a healthy environment ready for ample development, a "terrestrial

"Designing Carolina: The Construction of an Early American Social and Geographical Landscape, 1670–1719" (Ph.D. diss., William & Mary, 1998), 75; L.H. Roper, *Conceiving Carolina: Proprietors, Planters, and Plots, 1662–1729* (New York: Palgrave Macmillan, 2004), 26, 40; Aaron M. Shatzman, *Servants into Planters: The Origin of an American Image; Land Acquisition and Status Mobility in 17th-Century South Carolina* (New York: Garland Publishers, 1989), 3–28; Converse D. Clowse, *Economic Beginnings in Colonial South Carolina, 1670–1730* (Columbia: University of South Carolina Press, 1971), 56–67; M. Eugene Sirmans, *Colonial South Carolina: A Political History, 1663–1763* (Chapel Hill: University of North Carolina Press, 1966), 10–16; Robert M. Weir, *Colonial South Carolina: A History* (New York: KTO Press, 1983; reprint, Columbia: University of South Carolina Press, 1997), 53–55; Wood, *Black Majority*, 13–18; quote: Edelson, *Plantation Enterprise*, 35; Gary Livingston Hewitt, "Expansion and Improvement: Land, People, and Politics in South Carolina and Georgia, 1690–1745" (Ph.D. diss., Princeton, 1996), 12–13, 29–31; David L. Coon, "The Development of Market Agriculture in South Carolina, 1670–1785" (Ph.D. diss., University of Illinois at Urbana-Champaign, 1972), chapter 3.

[8] Joseph West to Lord Ashley, 2 March 1671, *Shaftesbury Papers*, 267.

[9] Joseph West to Sir G. Carteret, 2 March 1671, *Shaftesbury Papers*, 269; Joseph Dalton to Lord Ashley, 20 January 1671, *Shaftesbury Papers*, 376. Charles Town was officially changed to Charleston during the city's incorporation in 1783. I maintain modern spellings of the city unless historically relevant to state the original spelling.

[10] "An Old Letter," ca.March 1671, *Shaftesbury Papers*, 309.

paradise" or a "natural garden." Promoting Carolina, these tracts played on Europeans' biblical understanding of the world. Before facing the reality of the natural environment, newly arriving settlers had created a "paradise myth" of the Lowcountry. Believing that an "earthly paradise lay somewhere to the west" of Europe, colonists saw the "unaltered" landscape as a way to fulfill God's will for building a "new Arcadia."[11]

Although these promotional tracts team with inaccuracies from absentee authors motivated by the possibility of commerce, the descriptions of the Carolina climate, topography, and agriculture reveal Europeans' landscape desires, or, at the very least, appealed to the readers' prejudices. Seventeenth-century colonists promoted Carolina as a mild environment. Maurice Matthews wrote in 1680 that Carolina was "generally verry [*sic*] healthfull [and] it being a rare thing to hear of anybodies death." He optimistically, or deceivingly, claimed, "[s]ome years about July and August wee [*sic*] have the fevar [*sic*] and ague among us, but it is not mortal."[12] The air was "serene and exceedingly pleasant, and very healthy in its natural Temperament."[13] One French Huguenot, attempting to persuade future immigrants, claimed Carolina was "a little warmer than Paris," but the colony is "where one feels very fit."[14] In accord with Proprietors' desires to attract immigrants with farming experience, tracts stressed the "fruitfulness" of the land. Soil was "fertile" and the "ground yields greater abundance" for agriculture, wilderness of "groves of Timber Trees" intermix with the "Savana's" to create a landscape "to compare Carolina to those pleasant Parks in England."[15] To some English settlers, early Carolina was "a garden [rather] than an

[11] Edelson, *Plantation Enterprise*, chapter 1: "Laying Claim to the Land," specifically 13–24; H. Roy Merrens and George D. Terry, "Dying in Paradise: Malaria, Mortality, and the Perceptual Environment in Colonial South Carolina," *Journal of Southern History* 50 (November 1984): 534–535; Greene, "Early South Carolina and the Psychology of British Colonization," in *Imperatives, Behaviors, and Identities: Essays in Early American Cultural History* (Charlottesville: University of Virginia Press, 1992), 93, 108; Roderick Nash, *Wilderness and the American Mind*, 3rd ed. (New Haven: Yale University Press, 1982), 25, 33, 40–41; William Cronon, "The Trouble with Wilderness; or, Getting Back to the Wrong Nature," in *Uncommon Ground: Rethinking the Human Place in Nature*, ed. William Cronon (New York: W.W. Norton & Co., 1995), 70–71.

[12] Maurice Matthews, "A Contemporary View of Carolina in 1680," *South Carolina Historical Magazine* 55 (July 1954): 157.

[13] John Archdale, *A New Description of That Fertile and Pleasant Province of Carolina* (London, 1707; Charleston: A.E. Miller, 1822), 13.

[14] Louis Thibou letter, September 20, 1683, Louis Thibou Papers, South Caroliniana Library, University of South Carolina (USC).

[15] *Carolina, Described More Fully Then Heretofore* (Dublin, 1684), 21.

untilled place," and they promoted a sublime vision of a "bowling alley, full of dainty brooks and rivers of running waters." To the seventeenth-century reader, these descriptions represented the encouraging prospects of a new life associated with land ownership.[16]

Before widespread rice culture initiated changes in low-lying wetlands, colonial inhabitants first transformed the sandy loam. Seventeenth-century Europeans and their enslaved laborers settled the highest terrain, only five to ten feet above the mean high water mark in modern Charleston County, located in close proximity to navigable waterways. Once colonists claimed desirable tracts, subsequent immigrants traveled further upstream and inland. The free and enslaved initially lived within spatially compact settlements nestled on scarps and terraces that supported upland pine and oak communities. Early trade networks overlapped these ridgelines, as pathways and emerging turnpikes followed Indian paths, on terraces and highland conformities, to Charles Town. The sandy loam environments also supported "English grain," like barley and wheat, and experimental crops like cotton and tobacco. To European colonists, the elevated ridges became areas where they could recreate the pastoral landscapes of their homelands.[17]

During the first decades of colonization, settlers' approach to altering the Lowcountry landscape was based on an uncertain supply of labor. Property owners who arrived in Carolina with little capital were unable to purchase enslaved Africans or Native Americans. Landowners were

[16] Archdale quoted in Greene, "Psychology of Colonization," 93; Coon, "Development of Market Agriculture," 111–112; Richard I. Groening, Jr., "The Rice Landscape in South Carolina: Valuation, Technology, and Historical Periodization" (M.A. thesis, University of South Carolina, 1998), 35–38; Thibou letter, USC. Nash states during the pioneer period of settlement, when "nature" or wildness scenery did appeal to colonists, "it was not for its wilderness but because it resembled a 'Garden or Orchard in England.'" See: Nash, *Wilderness*, 33.

[17] Quote: John Stewart, "Letters from John Stewart to William Dunlop," *South Carolina Historical and Genealogical Magazine* 32 (January 1931): 16; Stanley South and Michael Hartley, *Deep Water and High Ground: Seventeenth Century Low Country Settlement*, Research Manuscript Series 166 (Columbia: University of South Carolina Institute of Archaeology and Anthropology, 1980), 4–6, 24–35; John Norris, *Profitable Advice for Rich and Poor, in a Dialogue, or Discourse between James Freeman, a Carolina Planter, and Simon Question, a West-Country Farmer, Containing a Description, or True Relation of South Carolina* (London: F. How, 1712), 36–41, describes the process of clearing land and plantation settlement in the early Carolina frontier. Anthropologists Stanley South and Michael Hartley point out that several of the late seventeenth-century settlements located on high ground, specifically on Johns Island, in relation to the North Edisto and Stono Rivers still cater to agricultural zones. See: South and Hartley, *Deep Water and High Ground*, 30.

inspired to initiate economic ventures that required little labor. Once Euro-American planters produced commodities for a world market, the Lowcountry landscape dramatically changed to reveal the human imprint of technology and society. Yet by the turn of the eighteenth century, Carolina's close association with the West Indian plantation complex set the colony apart from other North American colonies. Merchants established trade networks between Carolina, the Caribbean, and Great Britain. West Indian plantations' need for foodstuffs provided stimulus for Lowcountry colonists' agricultural ventures.[18] By 1690, Carolinians were exporting deerskins, naval stores, lumber, and salted meat to England and her colonies.[19]

Livestock ranching became a precursor to colonial South Carolina rice cultivation. As historian John S. Otto explained, "drawing upon British and African antecedents, cattle-ranching proved the ideal industry for early Carolina – a colony with an abundance of land and cattle but a shortage of capital and labor."[20] Cattle ranching took place on the three ecosystems that later would become habitats for rice: upland longleaf pine communities, small stream floodplains, and low-lying hardwood bottomlands. As colonists and slaves tended cattle, they became familiar with these ecosystems. Knowledge of these environments – partially obtained through cattle ranching – became a critical component for successful rice farming. Cattle ranching also increased Carolina's dependence on

[18] Russell R. Menard, "Slavery, Economic Growth, and Revolutionary Ideology in the South Carolina Lowcountry," in *The Economy of Early North America: The Revolutionary Period, 1763–1790*, eds. Ronald Hoffman et al., (Charlottesville: University Press of Virginia, 1988), 259–262, quote: 261–262; Sirmans, *Colonial South Carolina*, 7; Wood, *Black Majority*, 13–34; Gray, *Agriculture in the Southern United States*, 325; Emma Hart, *Building Charleston: Town and Society in the Eighteenth-Century British Atlantic World* (Charlottesville: University of Virginia Press, 2010), 17–37; Edelson, *Plantation Enterprise*, 43; Kinloch Bull, "Barbadian Settlers in Early Carolina: Historical Notes," *South Carolina Historical Magazine* 96 (October 1995): 336–339; Roper, *Conceiving Carolina*, 25–28; Shatzman, *Servants into Planters*, 3–28; Richard S. Dunn, *Sugar and Slaves: The Rise of the Planter Class in the English West Indies, 1624–1713* (Chapel Hill: University of North Carolina Press, 1972), 113; Hewitt, "Expansion and Improvement," 29–30.

[19] Russell R. Menard, "Economic and Social Development of the South," in *The Cambridge Economic History of the United States*, vol. 1: *The Colonial Era*, eds. Stanley L. Engerman and Robert E. Gallman (New York: Cambridge University Press, 1996), 275; Robert L. Meriwether, *The Expansion of South Carolina, 1729–1765* (Kingsport, TN: Southern Publishers, 1940), 4–5; Coon, "Development of Market Agriculture," 85–86.

[20] John S. Otto, "The Origins of Cattle-Ranching in Colonial South Carolina, 1670–1715," *South Carolina Historical Magazine* 87 (April 1986): 122.

enslaved labor, as colonists could afford to invest in more labor from their profits. Packed meat exported to the British West Indies became an early route to wealth and landholdings. Like the Indian trade, ranching required relatively little labor and capital. Colonists let their livestock range freely throughout the emerging plantation landscape; abundant land eliminated the need to construct fences and produce fodder. Hogs and cattle foraged freely "at no cost whatever" in the upland forests and savannas during the summer while feeding in hardwood bottomlands and marshland canebrakes during the winter. By the early eighteenth century, hogs were "in abundance" throughout the Lower Coastal Plain as "they go daily to feed in the Woods, where they rove several Miles feeding on Nuts and Roots."[21]

While livestock foraged through the three Lowcountry ecosystems freely, Carolina settlers found the longleaf pine communities particularly conducive to raising cattle. According to one early eighteenth-century traveler, the longleaf pine forests were "exceedingly good for a stock of cattle, and on which [planters] frequently settle their cow-pens."[22] The complex layering of the longleaf pine canopies mixed with an understory of grasses created savannas that, according to environmental historian Albert Way, had "an aesthetic of parklike openness."[23] Colonists continued the Native American custom of "carving" savannas out of upland pine forests by burning the understory grasses to hunt game and clear agricultural land. This human practice mimicked the natural phenomenon of lightning storms igniting the longleaf pine forests, leading to an

[21] Peter H. Wood, "They Understand Their Business Well: West Africans in Early South Carolina," in *Grass Roots: African Origins of an American Art*, eds. Dale Rosengarten, Theodore Rosengarten, and Enid Schildkrout (Seattle: University of Washington Press, 2008), 87–89; Stephen Pyne, *Vestal Fire: An Environmental History, Told through Fire, of Europe and Europe's Encounter with the World* (Seattle: University of Washington Press, 1997), 466, quote: 477; John S. Otto, "Livestock-Raising in Early South Carolina, 1670–1700: Prelude to the Early Rice Plantation Economy," *Agricultural History* 61 (1987): 13–24; Mart A. Stewart, "'Whether Wast, Deodand, or Stray': Cattle, Culture, and the Environment in Early Georgia," *Agricultural History* 65 (Summer 1991): 1–28; Thibou letter, September 20, 1683, USC; John Oldmixon, *The British Empire in America: Containing the History of Discovery, Settlement, Progress and State of the British Colonies on the Continent and the Islands of Americas*, vol. I, 2nd ed. (London: J. Brotherton et al., 1741), 521.

[22] "A Gentleman's Account of His Travels, 1733–34," in *The Colonial South Carolina Scene: Contemporary Views, 1697–1774*, ed. H. Roy Merrens (Columbia: University of South Carolina Press, 1977), 119.

[23] Albert G. Way, *Conserving Southern Longleaf: Herbert Stoddard and the Rise of Ecological Management* (Athens: University of Georgia Press, 2011), 11.

evolution of fire-dependent ecosystems. By manipulating these burnings, humans turned a natural phenomenon that evolved over millennia into a tool for their own benefit. Despite the introduction of humans into this equation, the longleaf communities still thrived with the growth of fire-adapted vegetation and the animals that fed on these species.[24]

Cattle ranching was instrumental in establishing the Carolina plantation enterprise as land and labor needed for this regime provided the foundation for later commodities. Salted beef and pork became the first successful venture in the colony and was the fourth most exported commodity, behind rice, deerskins, and indigo, by the mid-eighteenth century. Satisfying the demands abroad for salted beef and pork, ranching began bringing profits to Carolinians by 1682. As trade and herds increased, so did colonists' demand for more slaves. By 1708, at least 1,000 of the 1,800 enslaved Africans in South Carolina worked in the cattle industry. Ranchers needed large estates to feed the livestock adequately through this "wild cattle" method. One cow required fifteen acres to adequately graze. Some entrepreneurs who amassed more than 300 head of cattle began purchasing larger plantations to accommodate their livestock. Carolinians continued the English West Indies tradition of naming landscapes after cattle activity. Coincidently, these areas became important inland rice zones. Select plantations that supported the livestock industry contained desirable ecosystems to grow rice. "Cow Savannah," "Hog Swamp," and "Horse Savannah" reflect three low-lying landscapes west of the Ashley River where planters successfully cultivated the crop in the eighteenth century.[25]

[24] Way, *Conserving Southern Longleaf*, 7–12; Richard D. Porcher and Douglas A. Rayner, *A Guide to the Wildflowers of South Carolina* (Columbia: University of South Carolina Press, 2001), 91–92. South Carolina colonists also labeled upland savannas "Indian Old Fields," acknowledging the Native American presence on these spatial landscapes: see Hennig Cohen, "'Old Field School,' 'Cornfield School,' and 'Indian Old Field,'" *American Speech* 29 (October 1954): 225–226. A cow pen was not only "a pen or an enclosure but was also used to designate a large grazing area, usually between 100 to 400 acres in size": see Gary S. Dunbar, "A Southern Geographical Word List," *American Speech* 36 (December 1961): 294. For a general account of savannas, see Gordon G. Whitney, *From Coastal Wilderness to Fruited Plain: A History of Environmental Change in Temperate North America from 1500 to the Present* (Cambridge, UK: Cambridge University Press, 1994), 93–97.

[25] Coclanis, *Shadow of a Dream*, table 3–12: "Charleston, SC Exports November 1, 1747 – November 1, 1748," 80–81; Otto, "Livestock-Raising in Early South Carolina," 13–16; Ralph W. Tiner, *In Search of Swampland: A Wetland Sourcebook and Field Guide* (New Brunswick: Rutgers University Press, 1998), 57; Weir, *Colonial South Carolina*, 142; Henry A.M. Smith, "The Baronies of South Carolina: Landgrave Ketelby's Barony,"

Grazing lands also catered to the early development of rice cultivation and would become the conduit between the two commercial enterprises by the end of the seventeenth century. John S. Otto implies this connection, stating "planters cultivated rice in the 'low moist Lands' along rivers, and they grazed stock in the surrounding woods." Large property holdings, available capital, and enslaved labor, attained through the success of the livestock industry, were three elements that benefited aspiring rice planters. Otto explains that livestock ranches were a "prelude to the rice plantation economy." For example, Joseph Wigfall's early eighteenth-century shift from cattle ranching to rice provides an example of these broader changes in land use. A butcher by trade, Wigfall originally raised cattle on a 1,500 acre tract located on the western branch of Awendaw Creek and sold his butchered meat at Charles Town Beef Market on the northeast corner of Meeting and Broad streets. For the 1708 Christ Church Parish boundary, a surveyor used Wigfall's cow pens as a marker of delineation. The same year, Wigfall and his brother-in-law David Maybank split the property. Wigfall used the northern "Willow Hall" tract to graze cattle, while Maybank cultivated his 500-acre "Owendaw" tract, which was later renamed "Rice Hope."[26] By 1712, rice farming surpassed livestock ranching as the leading agricultural activity. That year Carolina exported 12,727 barrels of rice, valued at approximately £40,000 sterling compared to 1,863 barrels of salted beef and 1,241 barrels of salted pork, with a combined approximate value of £10,000

South Carolina Historical and Genealogical Magazine 15 (October 1914): 155; S. Max Edelson, "Planting the Lowcountry: Agricultural Enterprise and Economic Experience in the Lower South, 1695–1785" (Ph.D. diss., Johns Hopkins University, 1998), 51–52. Oldmixon noted in 1708 "about 40 years ago it was reckoned a great deal to have three to four cows, now some people have 1000 Head, and for one Man to have 200 is very common." Oldmixon, *The British Empire in America*, 520.

[26] Otto, "Livestock-Raising in Early South Carolina," 22–23, 24; Thomas R. Wheaton et al., *Archaeological Site Testing of Willow Hall and Walnut Grove Plantations, Francis Marion National Forest*, Francis Marion and Sumter National Forests Cultural Resource Management Report 92-13 (Stone Mountain, GA: New South Associates, April 22, 1992), 28–29; Alston Deas, "Wigfall of Christ Church Parish," Wigfall Family Papers, College of Charleston Special Collections, Charleston, SC, 1; Sir John Colleton sued Joseph Wigfall in 1706 for refusing to pay for fourteen steers that Wigfall purchased from Colleton the previous year. The judgment roll lists Wigfall as a "butcher" in St. Thomas and St. Denis Parish. See: George D. Terry, "'Champaign Country': A Social History of an Eighteenth Century Lowcountry Parish in South Carolina, St. Johns Berkeley County" (Ph.D. diss., University of South Carolina, 1981), 85–86. For information on the Charleston Beef Market, see: Martha A. Zierden and Elizabeth J. Reitz, *Archaeology at City Hall: Charleston's Colonial Beef Market*, Archaeological Contributions 34 (Charleston: The Charleston Museum, 2005).

sterling. By 1725, the Wigfalls shifted to growing rice on an Awendaw Creek tributary. Twenty-one enslaved laborers grew 725 bushels of rough rice while tending 220 head of cattle at Willow Hall. Joseph's brother Samuel reflected the transition between economic ventures, as he was simultaneously listed as a "planter" and a "livestock raiser," when owning the plantation in 1725. Representing the transition from cattle to rice enterprises, Wigfall and his descendants would continue to grow the grain on this property for the next 150 years.[27]

Cattle ranching gave enslaved Africans access to the diverse Lowcountry landscape. Part of enslaved cattle-hands' duties were to round up freely ranging livestock for the evening. Cattle and hogs foraged through the sprawling landscape, roaming through tidal marshes, upland savannas, and bottomland floodplains. Europeans hesitantly ventured into low-lying swamps; the task was left to enslaved Africans to tend to foraging livestock. In doing so, slaves familiarized themselves with the various Lowcountry ecosystems. In 1708, one writer noted the majority of slaves in Carolina "knows the Swamps and Woods, most of them Cattle-hunters." While planters attempted to define boundaries between plantations and the wilderness, slaves served as the "middling" between two environments, as S. Max Edelson explains. Everyday exposure to the environment enabled these people to put the landscape to work for their own benefit. Either actively herding animals for their masters or temporarily escaping into the wilderness for a brief reprieve, the early cattle-hands moved easily between the pineland savannas and the cypress bottomlands.[28]

By the turn of the eighteenth century, Carolina settlers' perceptions of a New Eden gave way to reality. Colonists first began to experience the affects of menacing weather. "To tell you the truth," confided one Huguenot émigré, "this country is not at all like it was depicted." The colony is good for those "who are resolved to suffer." Promotional

[27] Otto, "Livestock-Raising in Early South Carolina," 23; Deas, "Wigfall of Christ Church Parish," 4; Wheaton et al., *Archaeological Site Testing of Willow Hall and Walnut Grove Plantations*, 44.

[28] Otto, "Livestock-Raising in Early South Carolina," 15–20; quote: Clarence L. Ver Steeg, *Origins of a Southern Mosaic: Studies of Early Carolina and Georgia* (Athens: University of Georgia Press, 1975), 106; S. Max Edelson, "The Nature of Slavery: Environmental Disorder and Slave Agency in Colonial South Carolina," in *Cultures and Identities in Colonial British America*, eds. Robert Olwell and Alan Tully (Baltimore: Johns Hopkins University Press, 2006), 22, 24, 27. For African connections to early South Carolina cattle ranching, see: Andrew Sluyter, *Black Ranching Frontiers: African Cattle Herders of the Atlantic World, 1500–1900* (New Haven: Yale University Press, 2012), 136–138.

literature presented "only ... the good side and hardly ever talks about the difficulties that one endures in establishing oneself." With people trying to make sense out of an unknown country, they assessed healthy places based on sight and smell. The sultry temperatures became an indicator of poor health, as colonists attributed the heat with sickness and death. Colonists died from "exhaustion" when working in the heat and high humidity. Missionary Francis LeJeu described "the greatest danger" near Goose Creek "is to ride in the heat of ye day which is sometimes very great." He attributed Carolina's extreme temperatures in 1704 to "killing" a fellow missionary.[29]

Lowcountry colonists also witnessed disease epidemics from the beginning of June to the end of October. Recounting in 1687 how two former colonists "have never before seen so miserable of a country, nor an atmosphere so unhealthy," a Bostonian described Carolina with "fevers prevail ing] all the year, from which those who are attacked seldom recover." In 1682, 1684, and 1687, there were three notable seasons of disease epidemics, feeding on increasing immigration and wet summers.[30] An observer wrote in 1684, "who in this Country have seated themselves near great Marshes, are subject to Agues, as those who are so seated in England."[31] The summer of 1687 "was rather severe," according to a Santee resident, "with almost continuous rains and fevers that were commonplace." As colonists occupied land bordering the Ashley and Cooper Rivers, disease took its toll on the population. H. Roy Merrens and George D. Terry observed, "in some parts of the colony the mortality rate was so high that a number of parishes did not experience a natural increase in population until the American Revolution." Unaware that people introduced diseases – as well as some of the vectors – that thrived in part because of human landscape transformations, Lowcountry colonists made the basic connection that wetland environments were a death

[29] Nash, *Wilderness*, 25; Conevery Bolton Valencius, *The Health of the Country: How American Settlers Understood Themselves and Their Land* (New York: Basic Books, 2002), 23–43; Molly McClain and Alessa Ellefson, "A Letter from Carolina, 1688: French Huguenots in the New World," *William and Mary Quarterly*, 3rd ser. 64 (April 2007): 390, 394; Francis LeJau to Mr. Stubbs, July 3, 1704, The Society for the Propagation of the Gospel in Foreign Parts, A3: 266.

[30] Wood, *Black Majority*, 67; Timothy Silver, *A New Face on the Countryside: Indians, Colonists, and Slaves in South Atlantic Forests, 1500–1800* (Cambridge, UK: Cambridge University Press, 1990), 155–161.

[31] *Carolina, Described More Fully*, 20.

sentence to many inhabitants.[32] Governor Archdale pronounced at first that "Planters experimented, seldom having any raging sickness but what has been brought from the Southern Colonies," yet by 1707 he warned, "the late Sickness may intimidate" prospective colonists.[33]

Carolinians' views of wetlands reflected broader English perceptions of the low-lying ecosystems at the turn of the eighteenth century. Settlers in the new environment saw cypress and hardwood bottomland wetlands as "wastes," land "as unusable while still allowing for the kind of promise of a use not yet found" because they approached landscapes with European ideologies of land use.[34] Europeans attempting to construct their Eden viewed wooded wetlands as evil or "dismal." The dense impenetrable landscape, according to Ann Vilesis, "violated [seventeenth century colonists'] norms of orderliness and presented an incomprehensible, chaotic landscape – in contrast with the familiar English countryside and pastoral landscape that they sought to recreate."[35] Colonists idealized romantic old-world pastoralism and attempted to apply these sensibilities in Carolina as seen in the parklike metaphors used by promoters at the turn of the eighteenth century that reflected the ideal of an orderly and tamed landscape.[36]

Colonists first experimented with rice in upland environments near the Ashley and Cooper Rivers. Pine communities satisfied common English perceptions of the landscape in terms of health and value. Also, Carolinians' early practice of rice cultivation resembled their understanding of old-world farming. A 1666 survey of potential agricultural lands in Carolina listed rice as one of many grains that settlers could grow in the "meadows" of longleaf pine ecosystems. Biologist Richard Porcher notes that the savanna's limited tree cover made these landscapes easier to convert into agricultural zones without a large labor force. The clay lens approximately one foot underneath these microenvironments created moist soil for growing crops. Many English farming practices transferred

32 McClain and Ellefson, "Letter from Carolina," 382, 388; Merrens and Terry, "Dying in Paradise," 540–541, quote: 541; St. Julien Ravenel Childs, *Malaria and the Colonization of the Lowcountry, 1526–1696* (Baltimore: Johns Hopkins University Press, 1940), 33, 128, 158; Peter McCandless, *Slavery, Disease, and Suffering in the Southern Lowcountry* (New York: Cambridge University Press, 2011), 45–46.

33 Archdale, *A New Description*, 13.

34 Edelson, "Planting the Lowcountry," 56, quote: 58.

35 Ann Vilesis, *Discovering the Unknown Landscape: A History of America's Wetlands* (Washington, D.C.: Island Press, 1997) 33; Cronon, "Trouble with Wilderness," 70–71.

36 Edelson, *Plantation Enterprise*, 6.

to the New World could be adapted to this landscape, as planters transformed cleared mixed hardwood and pine forests into fields and constructed shallow ditches to drain moist savanna soil. At the encouragement of the Lord Proprietors, colonists incorporated rice and other crops into their planting schedule. John Stewart wrote in 1690 that he and his neighbors on the Cooper River were "bettering of all Kinds of English grain by sowing 3 bushels to an acre and 3 tymes [*sic*] pleug [plowing] first for Ryce[*sic*] barley and wheat." He noted how colonists experimented with growing grains in different soil types and that "the discovery of pine land to excel far out our oakground either for graine Englysh [*sic*] or Ryce [*sic*]." The same year, Stewart reported that Governor James Colleton devoted savanna land to cultivating rice, barley, wheat, peas, cotton, indigo, and Indian corn.[37]

Planters learned, either from their own experiments or from slaves, which crops worked well in which environments. For example, peas and corn could grow successfully in slightly higher soil in close proximity to rice. Even as colonists experimented with agriculture and harvested timber, they had to gain a new understanding of soil. John Steward advised Thomas Shepard that South Carolina upland soil "is lighter, sandyer dryer hotter and the sun hotter here than in Eurpo[sic] our gras [s] and weeds quicker in growth and our grain swells more than in [B]ritain when it is thin sowen."[38] With cultivation zones differing by a few feet (if not inches), variations in soil content contributed to specific crop location. Soil water permeability and physical properties changed in relation to topography. Unfamiliar with the new landscape, both planters and slaves had to devote time and keen observation to connecting specific settings with each crop. Through trial and error, planters observed which soil produced the best results for growing rice. While colonists first grew rice on dry highlands to a limited degree, they gradually realized that specific environments with moist soil produced higher yields.

Early Carolinians first grew rice in a "providence" style on savannas and nearby upland sites. Historians tend to label this general cultivation

[37] Clowse, *Economic Beginnings*, 125–6; "A Brief Description of the Province of Carolina, by Robert Horne (?), 1666," in *Narratives of Early Carolina, 1650–1708*, ed. Alexander S. Salley, Jr. (New York: Charles Scribner's Sons, 1911), 69; Richard D. Porcher, and William Robert Judd, *The Market Preparation of Carolina Rice* (Columbia: University of South Carolina Press, 2014) 30–34; Stewart, "Letters from John Stewart to William Dunlop," 16–17, 21–22.

[38] John Stewart, "Letters from John Stewart to William Dunlop," *South Carolina Historical and Genealogical Magazine* 32 (April 1931): 86.

method as "upland rice," yet Richard Porcher noted that this practice occurred in a variety of microenvironments and "is aptly called providential culture," for planters sowed seeds with "no provision for water control on the fields." Planting followed typical English farming practices of tilling soil and broadcasting seeds, and then the farmer hoped for rain to provide irrigation for the crop. In a style similar to sowing barley, planters cast rice seed in a thick cover which "chokt [*sic*] the weeds." Despite these transparent similarities, colonists soon realized that they could not duplicate old-world practices and that growing rice was "not like sowing of grain in England." Planters could not "put the plow in such land," as stumps and roots hindered initial tilling. The growing populations of enslaved laborers were forced to "take the trouble of digging the land with mattocks, hoes, or otherways, and sow English grain thereon." Also, as Philip Morgan explains, colonists abandoned broadcasting as slaves favored embedding rice seed into the soil, specifically by indenting the ground with one's heel and using the foot to slide soil over the seeds, a practice found in present day Mali and southern Benin.[39]

Just as early immigrants faced the reality of the Lowcountry climate in the wake of romantic Mediterranean associations, these colonists also experienced agricultural realities when intensively cultivating sandy soils. Colonists encountered problems with soil quality after making alterations to providence rice culture. Alexander Hewatt commented that the sandy highlands "poorly rewarded [the planter] for their toil." Soil exhaustion became apparent with the intensive farming practices upon the fine sandy loam. The endless land left colonists to disregard traditional English husbandry – seen in crop rotations, animal grazing, and periodic flooding on specific plots – in search for uncultivated property. Lowcountry colonists abandoned their English counterparts' careful maintenance of husbandry because of shortsighted practices removing key ingredients from mixed land use and greed to continually look toward more virgin land instead of nurturing existing cleared parcels. Early rice planters cultivated

[39] Stewart, "Letters from John Stewart to William Dunlop," 85; Porcher and Judd, *Market Preparation*, 32; Edelson, *Plantation Enterprise*, 103; Norris quoted in Merrens, ed., *Colonial South Carolina Scene*, 45, 46, 50; Morgan, *Slave Counterpoint*, 150–151; Henry C. Dethloff, "The Colonial Rice Trade," *Agricultural History* 56 (January 1982): 238; Edda L. Fields-Black, *Deep Roots: Rice Farmers in West Africa and the African Diaspora* (Bloomington: Indiana University Press, 2008), 159; Stanley B. Alpern, "Did Enslaved Africans Spark South Carolina's Eighteenth Century Rice Boom?," in *African Ethnobotany in the Americas*, eds. Robert Voeks and John Rashford (New York: Springer, 2013), 50.

a field three to four years before abandoning the plot, or "lay[ing] it out to grass," and clearing new land. Because planters allowed cattle to range freely throughout the Lowcountry, colonists could not effectively enclose cattle within abandoned fields without building surrounding fences. Without cattle naturally fertilizing the upland soil, fallow fields took longer to rejuvenate. By the time abandoned providence rice fields could be reintroduced into the rotation, rice cultivation had shifted to more fertile low-lying landscapes that maintained soil fertility for decades.[40]

Colonists continued to practice the providence culture during the last decade of the seventeenth century on small-stream floodplains, also called "dry swamps" or "oak and hickory land," that formed below the upland pine and savanna communities. Small-stream floodplains were localized alluvial watercourses, or first-order watersheds, providing the headwaters of Lowcountry tidal rivers. The vegetation of small-stream floodplains was "dominated by swamp trees with a herbaceous ground cover or canebreaks." The low or moderate soil permeability supported dominant species of black and sweet gum, live oak, red maple, and longleaf pine. Like the upland pine communities, small-stream floodplains had less groundcover compared to bottomland hardwood communities. To colonists, small-stream floodplain canopies covering the secondary grasses produced a pleasing park-like aesthetic that they could easily walk through, yet they provided the much-desired fertility to support the emerging rice economy.[41] A 1730s' account from a German Protestant settling in Georgia provides some insight into the subtle variation between small-stream floodplains and the hardwood bottomland:

> We are now learning to understand what the Englishmen mean when they said that swamps contained the best land. They do not mean swamps or bogs as we had in Ebenezer, which lie low, are always full of water and cannot be drained. Instead, they mean dry and low cane-covered regions and valleys in which water does not stand except when it is raining and from which it drains off quickly even

[40] Alexander Hewatt, *An Historical Account of the Rise and Progress of the Colonies of South Carolina and Georgia* (London: Alexander Donaldson, 1779), 158–159; Norris quoted in Merrens, ed., *Colonial South Carolina Scene*, 46; Carville Earle, "The Myth of the Southern Soil Miner: Macrohistory, Agricultural Innovation, and Environmental Change," in *The Ends of the Earth: Perspectives on Modern Environmental History*, ed. Donald Worster (New York: Cambridge University Press, 1988), 175–210.

[41] Porcher and Judd, *Market Preparation*, 30–34; quote: Stewart, "Letters from John Stewart to William Dunlop," 16, also, 17, 21–22; Catesby in Merrens, *Colonial South Carolina Scene*, 93; John D. Hodges, "Minor Alluvial Floodplains," in *Southern Forested Wetlands: Ecology and Management*, eds. Michael G. Messina and William H. Conner (Boca Raton, FL: Lewis Publishers, 1998) 325, 327–328.

then. Or they mean those in which nature has provided a small canal in which the water from the two hilly, cane-covered places can drain off. We have such swamps here, and everybody would like to have them.[42]

Small-stream floodplain soils were rich in nutrients, providing fertile microenvironments for agriculture with adequate moisture from surrounding streams and periodic freshets. Because small-stream floodplains often had a clay lens under the topsoil that retained surface water, draining practices were necessary for adequately cultivating crops. Soil in these floodplains was part of larger soil associations reflective of the watershed. The soil composition resulted from the length of the stream and how the floodplain fitted within the topographical setting. Early colonists conducting agricultural experiments on small-stream floodplains along the Ashley and Cooper Rivers generally grew crops in soil that modern scientists call Lenoir and Wahee loams. Both soil types presented suitable conditions for growing rice, with the higher elevation Lenoir fine sandy loam slightly more permeable compared to the lower elevation Wahee clay loam.[43]

As planters experimented with agriculture on small-stream floodplains, variations in soil and water encouraged them to incorporate a variety of agricultural practices. John Stewart served as Governor James Colleton's attorney while simultaneously managing Colleton's Wadboo estate along the headwaters of the Western Branch of the Cooper River. By 1690, he was attempting to grow mulberry trees, cotton, rice, wheat, barley, and grapevines on "22 severall [*sic*] places and differing grounds." Writing to William Dunlop, a former colonist then living in Scotland, Stewart noted that he made an "abundance of experiments on Rice Cotton and Silk which I promulgat evrywher [*sic*]." Stewart described growing rice and "English grains" at Wadboo Barony, emphasizing draining fields to remove standing water. He advised, "dry swamps to be clear'd before all other for vineyards and Ryce and with drains." The same year, Jean-Francois Gignilliat grew an identical list of "seeds" on his Santee River plantation where "every year successfully [I] try something that may

42 Quoted from Groening, "Rice Landscape," 72.

43 Hodges, "Minor Alluvial Floodplains," 325; Porcher and Judd, *Market Preparation*, 3; United States Department of Agriculture, *Soil Survey of Berkeley County, South Carolina* (Washington, D.C.: GPO, 1980), 19–20, 30–31, 95; Brooke V. James and Benjamin J. Collins, *Secondary Succession Patterns of a Southern Bottomland Hardwood Forest on Former Agricultural Lands in and around the Santee Experimental Forest, South Carolina* (Cordesville, SC: Santee Experimental Forest, United States Department of Agriculture, Forest Service Southern Research Station, 2010), 23.

make the country rich." Colonists imposed order on the landscape by straightening out meandering creeks and streams, while channels provided additional drainage when freshets inundated the crops. By the turn of the century, rice cultivators also began sowing seeds in furrows, or "trenches." Field hands would use approximately one to one and a half pecks of seeds per acre, "covering thin with earth," planting in rows twelve to eighteen inches apart between early April to mid-May. The furrow method used ten times less seed per acre compared to broadcasting.[44]

While Carolinians initially grew grains within an array of environments, the nutrient rich small-stream floodplains became desirable for growing rice. These floodplains provided the spatial transition or middle ground, both ecologically and culturally, between the upland pine barrens and the low-lying cypress hardwood floodplains. Stewart's letters to Dunlop reveal that planters were experimenting with lowland "swamps" by 1690. He points out that Colleton and neighboring planter Thomas Shepard were each attempting to drain swampland in order to plant grape vines and rice near Goose Creek. Their agricultural scale was small. Colleton devoted fifteen acres while Shepard planted four bushels of rice in "9 acres of swamp." Despite the two planters' attempts to grow rice in the moist soil, they were inundated with competing grasses and weeds that made cultivation "more drudgery ... than it was worth." Colleton, who followed "the Goosecreek philosophers' old measures," and Shepard were attempting to drain these wetland areas. These early Goose Creek planters still followed the "upland" method of cultivating rice by broadcasting seed and relying on rainfall and moisture-laden soil to nourish the crop. Late seventeenth-century planters encountered competition with unwanted vegetation suffocating the crop from sunlight and nutrients. Planters attempting to grow rice on the small-stream floodplains instructed their field hands to "hoe, weed, or cut up the grass, or other trash, growing between the said trenches and rice," up to three times in the summer.[45]

[44] Norris, *Profitable Advice*, 40; Oldmixon, *The British Empire in America*, 519; Stewart, "Letters from John Stewart to William Dunlop," 15–17, quotes: 16, 90; Gignilliat quoted in Bertrand Van Ruymbeke, *From New Babylon to Eden: The Huguenots and Their Migration to Colonial South Carolina* (Columbia: University of South Carolina Press, 2006), 207; Porcher and Judd, *Market Preparation*, 32–33.

[45] Stewart, "Letters from John Stewart to William Dunlop," 21–22, 84–86; Catesby in Merrens, *Colonial South Carolina Scene*, 93.

Draining swamps enabled South Carolina planters to cultivate more land, yet this practice did not single-handedly transform rice cultivation into an agricultural success. Lowcountry savannas, small-stream floodplains, and cypress bottomlands presented planters with drainage problems similar to those of English fens and flowing water meadows, yet rice irrigation required a more complex understanding of drawing water on and off the land.[46] For planters to cultivate rice on a commercial level, they had to increase their output and efficiency. Flooding rice fields enabled planters to begin this process. By the time John Norris wrote his 1712 promotional tract, planters had established a cycle of flooding their rice fields three times between April and September to eradicate weeds. Although water-driven milldams began appearing in South Carolina by the turn of the eighteenth century, this technology did not solve the complex method of drawing water onto the fields. To make that leap of committing to lowland watersheds and practicing routine flooding techniques, prospective rice planters had to look beyond English grain cultivation to impoundments and channels.[47]

Select enslaved Africans, on the other hand, possessed basic cultivation skills that observant Carolina planters incorporated with available European technology. One of the strongest arguments for advocates of the "Black Rice" thesis is that West Africans – unlike the English – possessed knowledge of "inflow" and "outflow" irrigation practices. In communities from Senegal to Benin, African cultivators developed a "rice knowledge system" that was "highly localized and specialized" to topographical conditions. West Africans developed diverse cultivation technologies, rice strains, tools, and agricultural language to cultivate specific topographies.[48] Just as South Carolina planters developed unique

[46] Edelson, *Plantation Enterprise*, 73–76; Groening, "Rice Landscapes," 60–65. Public drains commonly associated with fens did not take place on the Lowcountry landscape until the early eighteenth century. Inland rice planters started developing large-scale public drains that resemble the English drainage works beginning in the mid-eighteenth century, see Chapter 3.

[47] Stewart, "Letters from John Stewart to William Dunlop," 21–22, 84–86; Norris, *Profitable Advice*, 40; Carney, *Black Rice*, 103. Jean-Francois Gignilliat wrote in 1690 that planters were importing hand-powered stone mills from London to grind rice, wheat, barley, rye, and oats, as "there is only one water mill in all of Carolina." Robert Cohen and Myriam Yardeni, eds. "Un Suisse en Caroline du Sud à la fin du XVIIe siècle," translated by Harriott Cheves Leland, *Bulletin de la Société de l'Histoire du Protestantisme Français* 134 (1988): 8, manuscript on file at the Huguenot Society of South Carolina, Charleston, SC.

[48] Quotes: Carney, *Black Rice*, 58; Fields-Black, *Deep Roots*, 107–134, 156. Scholarship supporting the "Black Rice" thesis is supported by Wood, *Black Majority*; Littlefield,

inland irrigation systems relevant to local environments, so had generations of West African cultivators centuries before European contact. Rice cultivation practices developed along the Inland Delta of the upper Niger River in Mali some 2,000 to 3,000 years ago. Africans planted a domesticated rice grain, *Oryza glaberrima*, down the Niger River and throughout the inland and mangrove swamps along the West African coastline. By the time of contact with Portuguese explorers in the mid-fifteenth century, African communities had developed intensive irrigation techniques for growing subsistence rice.[49]

Although there are no surviving primary sources describing African technology transfer of rice cultivation to the Americas, four decades of research reveals that some enslaved Africans were extremely capable of using similar cultivation techniques for subsistence purposes in the New World. How knowledge transferred from enslaved Africans to their European enslavers is still questioned by scholars. Gregory E. O'Malley's

Rice and Slaves; Carney, *Black Rice*; Fields-Black, *Deep Roots*; Frederick C. Knight, *Working the Diaspora: The Impact of African Labor on the Anglo-American World, 1650–1850* (New York: New York University Press, 2010). Also: Gwendolyn Midlo Hall, "Africa and Africans in the African Diaspora: The Uses of Relational Databases," *American Historical Review* 115 (February 2010): 136–150. Scholarship questioning the "Black Rice" thesis is found in David Eltis, Philip Morgan, and David Richardson, "Agency and the Diaspora in Atlantic History: Reassessing the African Contribution to Rice Cultivation in the Americas," *American Historical Review* 112 (December 2007): 1329–1358; Edelson, *Plantation Enterprise*, 53–91; S. Max Edelson, "Beyond 'Black Rice': Reconstructing Material and Cultural Contexts for Early Plantation Agriculture," *American Historical Review* 115 (February 2010): 125–135. Walter Hawthorne, "From 'Black Rice' to 'Brown': Rethinking the History of Risiculture in the Seventeenth and Eighteenth Century Atlantic," *American Historical Review* 115 (February 2010): 151–163, offers a middle ground noting that both African and European influence played a significant role in the development of rice culture in North America. Further debate can be found in *Rice: A Global History*, eds. Francesca Bray et al. (New York: Cambridge University Press, 2015).

49 Edda L. Fields-Black, "Atlantic Rice and Rice Farmers: Rising from Debate, Engaging New Sources, Methods, and Modes of Inquiry, and Asking New Questions," *Atlantic Studies* 12 (September 2015): 282–284; Edda L. Fields-Black, "Untangling the Many Roots of West African Mangrove Rice Farming: Rice Technology in the Rio Nunez Region, Earliest Times to *c.* 1800," *Journal of African History* 49 (March 2008): 1–21; Olga F. Linares, "From Tidal Swamp to Inland Valley: On the Social Organization of Wet Cultivation among the Diola of Senegal," *Africa: Journal of the International African Institute* 51 (April 1981): 558, 560, 570–577; Michael Johnny, John Karimu, and Paul Richards, "Upland Swamp Rice Farming Systems in Sierra Leone: The Social Context of Technological Change," *Africa: Journal of the International African Institute* 51 (April 1981): 596–606; Olga F. Linares, "African Rice (*Oryza glaberrima*): History and Future Potential," *Proceedings of the National Academy of Sciences* 99 (2002): 16360–16365.

work on the transatlantic slave trade reveals the complexity (and pitfalls) of singling out specific ethnic groups related to African enslavement. Instead, he stresses that diversity prevailed in slave importation to the turn of the eighteenth-century Carolina. In order to challenge how we think of technology transfer as it relates to the creative adaptation of people in new landscapes, often relying on knowledge and creolization as it helps individuals specifically – a theme stressed in the work between Africans and Europeans – we must also recognize the complex layering of ideas between diverse African cultures as people were forced to make their way across the Atlantic and survive under brutal conditions in the New World. The creolization of enslaved Africans in the Carolina Lowcountry would more than likely occur with more fluidity compared to this acculturation between the enslaved and the enslavers, bound by a collective sense for survival. But the fact remains that free colonists transformed a subsistence culture into a commodified enterprise and there is a strong likelihood that they had African help.[50]

For West Africans transplanted into this new-world environment, nearby wetlands provided familiar zones for growing rice. "Africans modified their agricultural practices to suit the environment, rather than vice versa," according to Daniel Littlefield.[51] Relying on cultural memory, these enslaved cultivators brought a particular experience and sensibility to the landscape for agricultural experimentation. Africans' transfer of rice cultivation illustrates how cultural memory survived the Middle Passage and transplanted itself in the New World. Rice was a central part of the Rio-Nunez region, upper Guinea coast, Guinea-Bissau, and Gambian culture before colonial contact. As the slave trade connected West African ports to Barbados, enslaved Africans simultaneously transferred their cultural identity. Rice was part of their foodways, or how a culture expressed its food preferences through preparation. Rice was one of several staples transferred through the Middle Passage that later appeared in new-world gardens. Cereals (such as rice, millet, and sorghum), yams, black-eyed peas, sesame (benne), muskmelons, okra, and Guinea squash were all subsistence crops transferred from Africa to

[50] Gregory E. O'Malley, "Diversity in the Slave Trade to the Colonial Carolinas," in *Creating and Contesting Carolina: Proprietary Era Histories*, eds. Michelle LeMaster and Bradford J. Wood (Columbia: University of South Carolina Press, 2013), 234–255.

[51] Littlefield, *Rice and Slaves*, 86; Richard Price, "Subsistence on the Plantation Periphery: Crops, Cooking, and Labour among Eighteenth-Century Suriname Maroons," *Slavery and Abolition* 12 (no. 1, 1991): 107–127; Carney and Rosomoff, *In the Shadow of Slavery*, 148–154.

Carolina. Slave ship captains relied on these African staples to feed their enslaved cargo and keep them alive during the Middle Passage. Just as Africans formed a diaspora throughout the Lowcountry, so did the "shadow world of cultivation" that formed African subsistence diets.[52]

Rice's appearance in subsistence gardens coincided with the South Carolina colonists' period of experimentation, where perspective planters sought out plants that would take root in the fertile soil for both subsistence and profit. Free and enslaved farmers planted the African *O. glaberrima* and the Asian *O. sativa* strains in early Carolina. Ultimately, European markets and tastes preferred the white skinned *O. sativa*. By the late seventeenth century, English colonists were questioning how to incorporate rice into their diets and also have it serve as an exportable crop. John Stewart, for example, recommended substituting rice for barley to make beer and ale. Gignillait reported how rice served as an ingredient for eau-de-vie brandy. South Carolinian colonists incorporated rice into their staple diet, first by substituting ground rice flour for wheat and corn to simulate England's "fine wheaten bread" unavailable in the New World. The grain also provided additional "fodder" for poultry and livestock. Rice's versatility as a food for both Africans and Europeans, as historian Max Edelson explains, distinguished it from other plants grown for consumption and profit.[53]

Enslaved Africans' access to Lowcountry wetlands and small-stream floodplains allowed some of these individuals to practice subsistence agriculture. "On the plantation periphery," seventeenth-century slaves constructed rice fields in low-lying wetlands. Early plantation settlement patterns consisted of the planter's residence neighboring enslaved housing on terra firma knolls or ridgelines. Earthen swells, caused by Pleistocene deposits and resulting erosion, created a landscape surrounded by bays, streams, creeks, and rivers. The slaves' need to grow crops for survival challenged them to use land that free colonists considered undesirable.

[52] Fields-Black, "Atlantic rice," 282; Carney and Rosomoff, *In the Shadow of Slavery*, 124–125, 148–155.

[53] Carney and Rosomoff, *In the Shadow of Slavery*, 150–153; Fields-Black, "Atlantic Rice," 286–287; Edelson, "Nature of Slavery," 64–72; Stewart, "Letters from John Stewart to William Dunlop," 16; Cohen and Yardeni, "Un Suisse en Caroline du Sud," 8. David Coon points out the variety of seed rice found in colonial South Carolina, quoting John Lawson's 1709 report, "some bearded, others not, besides the red and white, but the white Rice is the best." Coon, "Market Agriculture," 169. For the significance of the color of commercial rice in nineteenth-century markets, see Bruce Mouser et al., "Red and White Rice in the Vicinity of Sierra Leone: Linked Histories of Slavery, Emancipation, and Seed Selection," in Bray et al., *Rice*.

They constructed embankments where they could grow patches of rice as they did in their homeland. Also, the presence of enslaved laborers in swamps, cutting timber or herding cattle, made them more acquainted with wetland hydrology. For personal reasons, Africans sought the plantation borderlands as a place of refuge. By removing themselves from the watchful eye of their enslavers, Africans used "down-time" to escape the oppression of slavery. As Peter Wood notes, these "black pioneers" were a mobile population that negotiated their way through swamps in tending to their duties. Slaves grew rice as one of a number of subsistence crops upon land that was unwanted by their masters – one of the many ways that Africans survived in the Lowcountry.[54]

Here lies the hybridized Lowcountry landscape – a synthesis of African, European, and Native American cultures – that still exists today. As colonists evaluated swamps for rice cultivation in the early eighteenth century, they became more optimistic about these environments and their productive potential. A movement toward low-lying landscapes came directly from people's desire to harvest timber and cultivate rice. By 1730, Europeans and Africans transformed the wetland landscapes from natural formations into suitable agricultural sites. Yet as Edelson notes, "planters learned to cast off the categories of conventional husbandry" in order to successfully cultivate rice.[55] Jack P. Greene attributes this change in landscape perceptions to the "psychology of colonization." For European colonists, reconstructing the environment symbolized "improved societies" and benefited their families and future generations. According to colonial-era naturalist Mark Catesby, "lofty trees of mighty bulk" prevented water from flowing through swamps, and cast dark hues by "excluding the sun's beams." Harvesting cypress, for instance, allegedly improved the swampland and "made the earth better adapted to the culture of rice." The wilderness was a disorderly and primitive environment that colonists had to alter. Colonists could provide order to wetlands by clearing land and channelizing water, as Greene notes, "through which the economic position of the nation might be advanced, the estates or fortunes of individuals bettered, or existing resources made more productive." For many of the aspiring planters, they had to venture

[54] Price, "Subsistence on the Plantation Periphery," 107–127; Wood, *Black Majority*, ch. 4: "Black Pioneers," 119–124.

[55] Edelson, *Plantation Enterprise*, ch. 2: "Rice Culture Origins," quote: 53.

out of the high-ground pastoral landscape and go into the low-lying wilderness in order to better themselves and future Euro-Americans.[56]

Once European colonists recognized the importance of impounding water to the irrigation of the rice crop and the simultaneously eradication of competing grass, a dramatic shift in landscape perceptions and in agricultural activity occurred. Rice farming moved from the upland and savannah ecosystems down to the cypress-hardwood stream systems. Colonists looked beyond the "wasteland" to see unlimited potential in transforming the low-lying small-stream floodplains, located "at the head of creeks and rivers," into orderly agricultural zones. The flow of water through wetlands fed the dense vegetation that created the apparently "inexhaustible fertility" of the South Carolina Lowcountry.[57] As the vegetation died and decomposed, nutrients accumulated and added to the soil's fertility that inland rice plantations exploited. Catesby noted that inland swamps were "impregnated by the Washings from the higher Lands, in a Series of Years are become vastly rich, and deep of Soyl [*sic*] consisting of a sandy Loam of a dark brown Colour."[58] One rice planter described inland swamps as having a "better foundation and soil than any other lands" and "by nature more durable" for cultivation because of the "fine supplies of decayed vegetable, which are deposited while the waters

[56] Greene, "Early South Carolina and the Psychology of British Colonization," 103–104; Catesby quoted in Merrens, *South Carolina Colonial Scene*, 93; Van Ruymbeke, *From New Babylon to Eden*, 205; Edelson, *Plantation Enterprise*, ch. 2: "Rice Culture Origins"; Terry, "Champaign Country," 48–89; Nash, *Wilderness*, 40–41; Carolyn Merchant, "Reinventing Eden: Western Culture as a Recovery Narrative," in *Uncommon Ground: Rethinking the Human Place in Nature*, ed. William Cronon (New York: W.W. Norton & Co., 1995), 132–159; Gary Hewitt states that South Carolina's agricultural transformation toward a plantation economy represents the process of social maturation, where "development of plantation agriculture was viewed as a sign of progress, to be encouraged by the colonial state as an indicator of civilization." Hewitt, "Expansion and Improvement," 60.

[57] "Reclamation of Southern Swamps," *DeBow's Review and Industrial Resources* 17 (November 1854): 525; Catesby quoted in Merrens, *Colonial South Carolina Scene*, 93.

[58] Catesby quoted in Merrens, *Colonial South Carolina Scene*, 92; noting the tremendous amount of physical labor needed to transform this environment into agricultural fields, Catesby commented that "this soil is composed of a blackish sandy loam, and provides good rice land, but the trouble of grubbing up and clearing it of the trees and underwood has been hitherto a discouragement to the culture of it," in Merrens, *Colonial South Carolina Scene*, 93; Wood, *Black Majority*, 59–62; Judith Carney, "Landscapes of Technology Transfer: Rice Cultivation and African Continuities," *Technology and Culture* 37 (Spring 1996): 14–16.

are passing over said lands."[59] These nutrient-rich soils provided European colonists the leeway of implementing intensive agricultural practices without suffering the consequences of poor husbandry.

Planters' success or failure began with selecting a site with the right soil content. Impounded rice fields best retained water in less permeable loam and clay in low-lying wetlands. Planters' and slaves' knowledge of soil content became crucial for the construction of reservoirs and fields. Although inland fields were localized in distinct watersheds, the microenvironments used for inland rice cultivation contained the same soil features. Meggett loam, as it was later known, was the soil type often associated with impounded inland rice zones. This soil had a mixture, or a loam, of sand, clay, and organic matter. These watersheds – or second-order stream-floodplains – fed into the Cooper-Ashley-Wando (CAW) Basin, the Ashepoo-Combahee-Edisto (ACE) Basin, and the Savannah River, among others, meandering throughout the South Carolina Coastal Plain.[60]

Slow water permeability and high water-holding capacity were two characteristics found in the rice fields. Slow water permeability means that the soil content prevents water from efficiently draining through the ground. This feature allowed rice cultivators to retain water in the reservoirs and the fields. Because of their low water permeability, soil from these zones was also used to construct the embankments. By reinforcing

[59] "Observations on the Winter Flowing of Rice Lands, in Reply to Mr. Munnerlyn's Answers to Queries, &c. by a Rice Planter," *Southern Agriculturalist and Register of Rural Affairs* 1 (December 1828): 531.

[60] Information about soil content derives from modern soil surveys, primary sources, and archaeological excavations. The Soil Conservation Service in the United States Department of Agriculture classifies, defines, and maps specific soils to current physical formations. Surprisingly, the physical remnants of inland and tidal rice fields still exist. Aerial photographs that contribute to the soil maps, or boundaries, for each unit reveal the inland fields and reservoirs. The soil content is a representation of what impact occurred on this land. As a result, the soil scientists acknowledge which soils were conducive for rice cultivation. For example, a 1916 soil survey explained that this soil type was "not used for agriculture, but abandoned canals, ditches, and dikes indicate that a considerable acreage was at one time used for the production of rice." The soil scientists observed that agricultural activity had not occurred for some time and that the land had become reclaimed "with a growth of cypress and gum, with longleaf and black pine, beech, and myrtle in the areas of slight elevation." United States Department of Agriculture: *Soil Survey of Beaufort and Jasper Counties* (Washington, D.C.: GPO, 1980), *Soil Survey of Charleston County* (Washington, D.C.: GPO, 1971), *Soil Survey of Berkeley County, South Carolina* (Washington, D.C.: GPO, 1980); quote: W.J. Latimer, ed. "Soil Survey of Berkeley County, South Carolina" in *Field Operations of the Bureau of Soils, 1916* (Washington, D.C.: Bureau of Soils, 1921), 515.

the retaining walls with clay, slaves created a basin to hold water within the natural terrain. However, the compacted soil that created desirable conditions for retaining water also created hardship for enslaved laborers shaping the landscape. South Carolina Governor James Glenn wrote that "the best land for rice is a wet, deep, miry soil; such as is generally to be found in Cypress Swamps; or a black greasy Mould with a Clay Foundation; but the very best lands may be meliorated by laying them under water at proper Seasons." One Wando River plantation overseer complained that his hoes were "too broad and soft to dig up the clay land." The hoes were useless "for digging up light clay land," as "they duble [*sic*] up like a sheet of lead and manney [*sic*] become useless before they are halfe wore [*sic*]." Inland plantations' clay loam posed difficulties in excavation, yet the soil provided an accessible medium for Africans and African Americans to express their cultural identity through making colonoware – a low-fired earthenware pottery – and architectural features revealed in clay-walled houses and chimneys.[61]

Planters' ability to draw a steady amount of water to the fields was the second characteristic needed to successfully cultivate inland rice. Consistent access to water in these cultivation zones enabled planters and slaves to grow rice in these inland settings. Unlike tidal rice cultivation where planters and slaves harnessed the "estuary hydraulics" of the river's ebb and flow, inland planters relied on water simply flowing from higher elevations down to their fields. These cultivators had to contain the natural resource from reliable surface and ground water sources, such as drainage basins, swamps, bays, and springs. Watersheds composed of Meggett loam were relatively level, so water flowed through these zones as a slow-moving current.[62]

[61] Glen quote: James Glen, *A Description of South Carolina: Containing Many Curious and Interesting Particulars Relating to the Civil, Natural and Commercial History of That Colony in Colonial South Carolina: Two Contemporary Descriptions*, ed. Chapman J. Milling (Columbia: University of South Carolina, 1951), 14; "Matthias Rash, Charleywood, to Peter Taylor, Whitehaven," 18 March 1773, Taylor Family Papers, 1709–1829, USC; Theresa A. Singleton, "Reclaiming the Gullah-Geechee Past: Archaeology of Slavery in Coastal Georgia," in *African American Life in the Georgia Lowcountry: The Atlantic World and the Gullah Geechee*, ed. Philip Morgan (Athens: University of Georgia Press, 2010), 157–170; Leland Ferguson, *Uncommon Ground: Archaeology and Early African America, 1650–1800* (Washington, D.C.: Smithsonian Institution Press, 1992), 18–22, 68–92. On the study of plantation hoes, see Chris Evans, "The Plantation Hoe: The Rise and Fall of an Atlantic Commodity, 1650–1850," *William and Mary Quarterly* 69 (January 2012): 71–100.

[62] Sam B. Hilliard, "Antebellum Tidewater Rice Culture in South Carolina and Georgia," in *European Settlement and Development in North America: Essays on Geographical*

The size and shape of inland rice fields were adaptations to topography. The basic inland rice field consisted of two earthen dams enclosing a low-lying area bordered by ridges. Enslaved people built up the "strong banks" with available fill from adjoining drainage trenches. The dam on the higher elevation contained stream or spring fed water to form a reservoir, or a "reserve," that would provide a water supply to the lower rice fields. Once cultivators released water from the reservoir, a second dam retained this resource to nourish rice fields. Between these two earthen structures was a series of smaller embankments and ditches to hold and drain water effectively during the cultivation process.[63]

Water control for inland rice cultivation required not only precise construction of earthen embankments, but also an understanding of the surrounding topography. Inland cultivators had to choose where to put reservoirs and fields in relation to watercourses and terrain. To retain water in the reservoir and rice fields, the soil required a substantial clay foundation to prevent impounded water from seeping out. The subtle elevation change, in some cases just three or four feet from sandy highlands to alluvial swamps, allowed different types of vegetation to take root. This diverse flora provided cultivators insight into soil composition. For example, longleaf pine and oak communities grew in well-drained sandy soil while cypress and tupelo gum communities grew in less permeable soil. To aspiring rice cultivators who did not have access to the insights of soil science until the middle of the nineteenth century, the distribution of trees and other plants directed them toward appropriate inland sites. Colonial-era historian Alexander Hewatt observed that "nature points out to [the planter] where to begin his labours; for the soil, however various, is every where easily distinguished, by the different kinds of trees which grow upon it."[64] In discussing how to locate "good

Change in Honour and Memory of Andrew Hill Clark, ed. James R. Gibson (Toronto: University of Toronto Press, 1978), 97.

63 Hilliard, "Antebellum Tidewater Rice Culture," 97; Hewatt, *Historical Account*, 303; Mart A. Stewart, *"What Nature Suffers to Groe": Life, Labor, and Landscape on the Georgia Coast, 1680–1920* (Athens: University of Georgia Press, 1996), 168.

64 Hewatt, *Historical Account*, 305; Joyce Chaplin notes although soil sciences did not evolve until the mid-1800s, people developed a basic understanding of which soils were fertile and which soil did not allow proper drainage. Joyce E. Chaplin, *An Anxious Pursuit: Agricultural Innovation and Modernity in the Lower South, 1730–1815* (Chapel Hill: University of North Carolina Press, 1993), 144–145; Porcher and Rayner, *Wildflowers of South Carolina*, 89–96; S. Max Edelson, "Clearing Swamps, Harvesting Forests: Trees and the Making of a Plantation Landscape in the Colonial South Carolina Lowcountry," *Agricultural History* 81 (Summer 2007): 386–390.

soil," Jean-Francois Gignilliat explained, "one recognizes [the soil] by the difference in trees, which are big oaks and nut trees of three or four types." Planters needed a "careful observance of topography and water flow" in selecting sites for rice cultivation. With substantial start-up costs to grow rice, planters had to understand hydrology and topography to avoid commercial failure before beginning the expensive endeavor.[65]

Topography determined the natural boundaries for reservoirs and fields, as high land enclosed plots and retained water. Elevation change from knolls to bottomlands helped enclose the reservoir and fields, eliminating the need for additional embankments. Because planters relied on geographical features to contain flooded fields, the boundaries of these agricultural systems initially resembled tributaries' fluid contours. Traveling through coastal South Carolina, Catesby observed how these inland landscapes took shape: "the further parts of these marshes from the sea, are confined by higher lands, covered with woods, through which by intervals, the marsh extends in narrow tracts higher up the country, and contracts gradually as the ground rises." Elevation differences between highlands and the rice fields ranged from four feet at Charleywood Plantation on the Wando River to forty feet at Newington Plantation located on the upper Ashley River. Compared to tidal rice fields, inland fields were smaller and contained within the topography of the wetlands. Tidal fields, on the other hand, sprawled across the riverbanks. The floodplains allowed planters to devote more land to tidal cultivation if the fields were built within the ebb and flow of the river.[66]

"Swamps had to be diked to separate land from water," observes Theodore Rosengarten, by enslaved people who, "cleared and chiseled [the floodplain] with hoes until it was as level as a billiard table." Dense hardwoods, such as bald cypress, tupelo, and sweet gum, were removed with axes and saws. John Norris observed in 1712 that stumps and roots took twelve to fifteen years to rot out of the fields, leaving slaves to plant rice around the remnant vegetation. Clearing the dense forests took an unimaginable amount of labor. Slashing and burning the fields expedited the decomposition process, as fire "softened" the landscape. Environmental historian Stephen Pyne notes, "with fire it was possible to reshape the

[65] Cohen and Yardeni, "Un Suisse en Caroline du Sud," 10; Carney, "Landscapes of Technology Transfer," 16.

[66] Catesby quoted in Merrens, *Colonial South Carolina Scene*, 92; elevation determined from current data referenced from USGS The National Map, http://nationalmap.gov/ (accessed April 8, 2019).

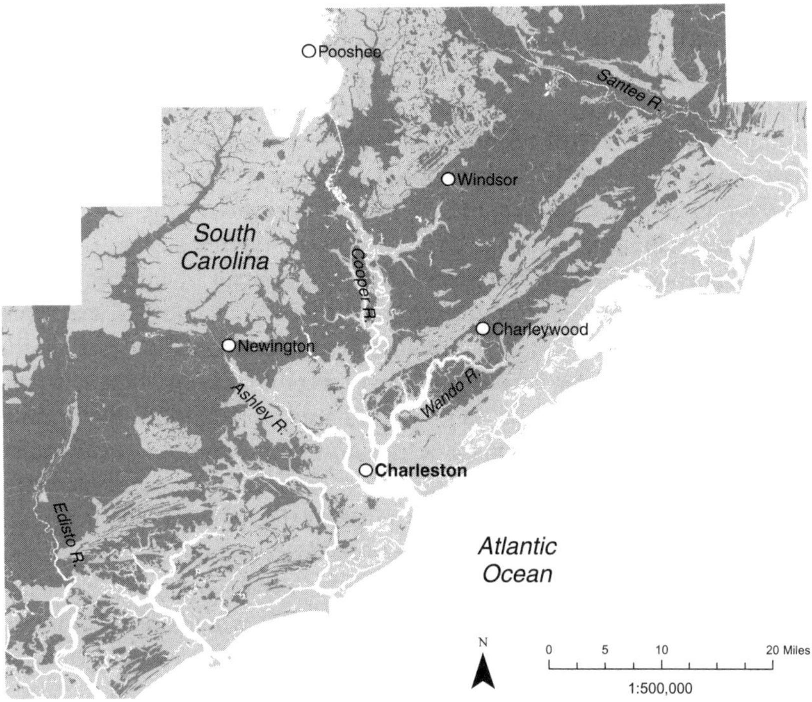

FIGURE 2.1 The Lower Coastal Plain between the Edisto and Santee rivers, South Carolina, with Pleistocene deposits highlighted in dark gray. Image adapted by author. South Carolina Department of Natural Resources, Geographic Information System (GIS) Division

pieces of the landscape mosaic and rearrange them into new pictures." Enslaved cultivators burned the underbrush and then hoed out the weed roots to prevent recurring growth of competing vegetation. Field hands spent January and February, "down" months in the agricultural cycle, burning leftover stubble on existing rice fields or clearing new acreage. Once vegetation was removed in South Carolina inland tracts, slaves leveled potential fields to accommodate rice planting and water drainage. After fields were developed to drain standing water, enslaved people dug precise quarter ditches to remove floodwaters more effectively. By the 1730s, such geometrically shaped fields had replaced the fluid landscape, redefining the non-human landforms of streams, banks, and knolls.[67]

[67] Theodore Rosengarten, "In the Master's Garden," in *Art and Landscape in Charleston and the Lowcountry*, ed. John Beardslay (Washington, D.C.: Spacemaker Press, 1998), 40; Carney,

The mechanical devices used to allow water on and off of the fields became a critical technical component for inland rice agriculture. Rice cultivators used gates, or "trunks," to control water flow from reservoirs onto the rice fields. Originally made from hollowed-out trees, trunks were traditional African devices found in the Rio-Nunez region used to regulate water flow through a conduit by plugging the end of *Borassus aethiopum* or *B. akeassii* palms. Enslaved Africans in the Lowcountry substituted domestic *Sabal* palms and cypress for this device. European colonists contributed to the process of water control by introducing the use of "valves" when constructing inland rice fields. Valves were rectangular boxes open on one end with a perpendicular sliding gate on the other to control water flow. These valves were used in draining fens and could serve a similar role on South Carolina tidal plantations; both trunks and valves were used to manage the downward flow of water while simultaneously preventing incoming tides from flowing onto the fields. Trunk engineers would eventually replace valves with a swinging gate design. Despite the swinging gate appearing later on tidal rice trunks, this design would have little benefit to impound reservoirs and inland fields. Inland cultivation depended upon technology that would allow control of water downstream, while the swinging gate's purpose was to prevent tidal surges from encroaching upstream into the fields.[68]

Black Rice, 86; Merrens, *Colonial South Carolina Scene*, 93; Littlefield, *Rice and Slaves*, 89; James M. Clifton, ed., *Life and Labor on Argyle Island: Letters and Documents of a Savannah Rice Plantation, 1833–1867* (Savannah, GA: Beehive Press, 1978), x. Period historian Alexander Hewatt speculated rice cultivation as "so laborious is the task of raising, beating, and clearing this article, that though it had been possible for attacking to the thick forest and clearing the grounds for the purpose, thousands and tens of thousands must have perished in the arduous attempt," in Hewatt, *Historical Account*, 120; Mathurin Guerin Gibbs Plantation Register, January 2, 1845, January 28–31, 1845, February 1–10 , 1845, South Carolina Historical Society (SCHS), Charleston, SC; Norris in Merrens, *Colonial South Carolina Scene*, 45; Pyne, *Vestal Fire*, 466, 468. Catesby believed that stumps would rot out in fields between six to eight years, in Merrens, *Colonial South Carolina Scene*, 93.

68 Littlefield, *Rice and Slaves*, 97; Carney, *Black Rice*, 95–96; Fields-Black, *Deep Roots*, 159; David Doar, *Rice and Rice Planting in the South Carolina Lowcountry* (Charleston: Charleston Museum, 1936; reprint, 1970), 12; Groening, "Rice Landscape in South Carolina," 158–159; Porcher and Judd, *Market Preparation*, 45–46. English farmers also used "hatches" to control water when floating meadows. Although hatches are associated with English technological transfer to Carolina rice fields, these devices did not resemble rice trunks in design. See: Eric Kerridge, *The Agricultural Revolution* (London: Allen & Unwin, 1967), ch. 6; Susanna Wade Martins and Tom Williamson. "Floated Water-Meadows in Norfolk: A Misplaced Innovation." *Agricultural History Review* 42 (1994): 21–30. For an image of an English hatch, see: George Boswell, *A Treatise on Watering Meadows* (London, 1779), plan no. 5.

Field engineers placed trunks in sloughs, or stream channels, so water efficiently ran out of the holding pond from the embankment's lowest point. Sloughs were an important natural feature for draining the wetlands, for they served as a "gutter," or a depression in the subtle elevation change.[69] After these floodwaters nourished the soil and rice crop, and killed competing weeds, the fields were drained through trunks located at the second embankment. The water released from these fields flowed downhill toward nearby tidal rivers.[70]

Inland rice planters were limited in terms of flood control by the single downward direction of water flow. Ironically by impounding water in reservoirs, planters actually created a precarious situation if freshets or droughts occurred. Freshets happened when storms or hurricanes provided more rain than the soil could absorb and streams could channel, causing "rapid torrents" that were "sudden and violent" as they flowed down hill.[71] This rush of water would flood the reservoir, causing an overflow that would breach the dam.[72] Droughts presented another problem for inland rice planters, as reservoirs that did not receive enough water would eventually dry up. Without plentiful "reserve water" to flood the fields, competing weeds overtook the orderly agricultural landscape.[73]

To better understand how people worked within the confines of the natural environment, three inland rice plantations – Pooshee, Newington, and Charleywood – provide examples representing how planters and their slaves designed rice cultivation strategies during the first quarter of the eighteenth century. By 1730, various forms of inland rice systems took shape along available streams throughout the South Carolina Lowcountry. Planters took advantage of diverse geographical features by integrating the basic inland model with subtle differences in field design, water control, and layout. Early eighteenth-century planters relied on small tributaries' definable floodplains to experiment with modes of irrigation control such as dams, embankments, ditches, and drains. In the French

69 Merrens, *Colonial South Carolina Scene*, 92.

70 Hilliard, "Antebellum Tidewater Rice Culture," 97; Carney, "Landscapes of Technology Transfer," 25.

71 Catesby quotes in Merrens, *Colonial South Carolina Scene*, 96.

72 Hilliard, "Antebellum Tidewater Rice Culture," 79; Sam B. Hilliard, "The Tidewater Rice Plantation: An Ingenious Adaptation to Nature," in *Geoscience and Man* 12, ed. H.J. Walker (Baton Rouge: Louisiana State Press, 1975), 58.

73 Stewart, "*What Nature Suffers to Groe*," 93; Chaplin, *An Anxious Pursuit*, 229–230; Clowse, *Economic Beginnings*, 127.

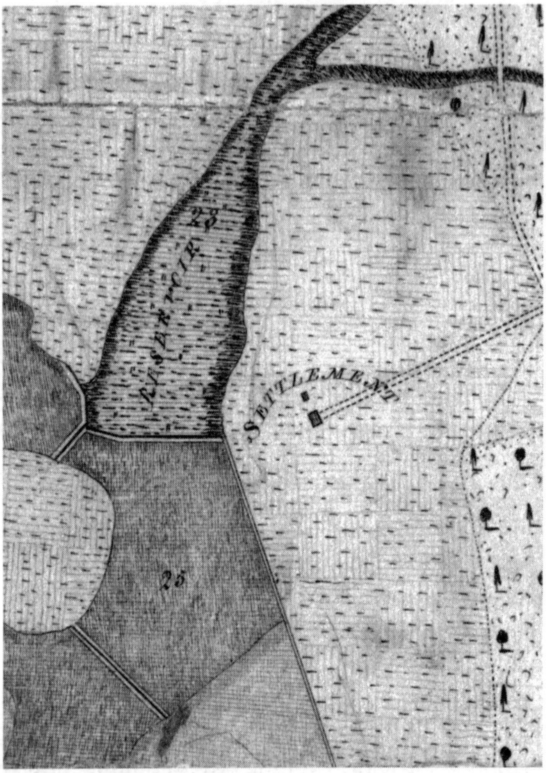

FIGURE 2.2 Simple inland rice field construction.
"Plan of Limerick, a Plantation Belonging to Elias Ball, esq." (detail), June 1797, John McCrady Plat Collection, 1541, Charleston County Register Mesne Conveyance, Charleston, SC

Huguenot enclave encompassing Biggin Swamp, at the headwaters of the Cooper River, neighboring planters tapped into the basin streams. Pooshee Plantation was one of thirty properties relying on Biggin Swamp tributaries for rice irrigation. The Lord Proprietors granted the 1,000 acres of the plantation to Peter St. Julien, a Vitré Brittany Huguenot, in May 1704. St. Julien sold the plantation to his brother-in-law, Henry LeNoble, in 1711, who three years later gifted Pooshee to his daughter Susanne and her new husband René Lewis Ravenel.[74]

Ravenel used a limestone spring, formed from a "downdip" in the Floridian aquifer system, to irrigate his Pooshee Plantation rice fields.

[74] "History of Transfer of Pooshee to 1756," Ravenel Land Papers, 1695–1880, South Carolina Historical Society; Van Ruymbeke, *From New Babylon to Eden*, 59, 86.

Occurring more frequently in the Penholoway Terrace, these artesian springs, or "fountains" as the local residents called them, provided consistent water flow for rice plantations throughout the Biggin Swamp community. Pooshee Springs was one of six notable fountains bordering the basin that established this area as one of the central rice zones in colonial South Carolina. Enslaved laborers at Pooshee, as S. Max Edelson notes, "made comparatively simple alterations to the land that took advantage of the existing contours of its topography." Slaves dug into Pooshee's "gray, sticky sandy clay" loam and "threw a dam" between the higher fine sandy loam to form a reservoir. Then they constructed a second dam to impound spring fed water and maintained the modest twelve-acre field.[75]

Newington Plantation, located on a tributary of the upper Ashley River, was an early eighteenth-century plantation located on the Penholoway Terrace. By 1701, the Axtell family was able to grow rice successfully for export using impounded water from Booshoo Creek. The creek curved around the plantation settlement's highland knoll, resembling a "C" shape. Enslaved Axtell laborers constructed a reservoir dam to hold back the creek. The reservoir powered the Axtell's sawmill while also flooding the downstream rice fields. What differentiated Newington from Pooshee's field system was the fact that Axtell slaves had to carve drains and embankments around the highland. Unlike Pooshee fields, which were irrigated by a spring-fed stream, Newington's fields were parallel, on the opposite side of the knoll, from Booshoo Creek. The channel followed the natural path of the creek and low land to drain into the seven-acre field below. The channeling effort paid off, as existing account

[75] Walter R. Aucott and Gary K. Speiran, "Ground-Water Flow in the Coastal Plain Aquifers of South Carolina," *Ground Water* 23 (November/December 1985): 738; George E. Siple, "Some Geologic and Hydrologic Factors Affecting Limestone Terraces of Tertiary Age in South Carolina," *Southeastern Geology* 2 (August 1960): 2; C. Wythe Cooke, *Geology of the Coastal Plain of South Carolina*, Bulletin 867 (Washington, D.C.: US Geological Survey, 1936), 75; Walter R. Aucott, "The Predevelopment Ground-Water Flow System and Hydraulic Characteristics of the Coastal Plain Aquifers of South Carolina," USGS Water Investigations Report B6–4347 (Washington, D.C.: GPO, 1988), 9; Edmund Ravenel, "The Limestone Springs of St. John's Berkeley," *Proceedings of the Elliott Society* (October 1860): 28–31; F.S. Holms, "Notes on the Geology of Charleston, SC," *American Journal of Science and Arts* 7 (March 1849): 665; "Map of Pooshee," 1920, Berkeley County Register of Mesne Conveyance, Monks Corner, SC (BCRMC); quotes: Edelson, *Plantation Enterprise*, 104, and *Soil Survey of Berkeley County*, 516, 506; "Rene Ravenel Inventory, 8 August 1750," Miscellaneous Inventories and Wills, Charleston County, 1687–1785, 77B: 578, South Carolina Department of Archives and History (SCDAH), Columbia, SC.

records reveal that the Axtell produced an average of ten thousand pounds of rice annually between 1701 and 1707.[76]

Toward the coast, the Princess Anne Terrace's brackish tidal rivers presented new challenges for early rice cultivators. Because of the terrace complex's close proximity to the ocean, Princess Anne began at sea level with a "gently inclined slope" up to twenty feet. The ocean's incoming tide pushed a "salt wedge" of brackish water against the downward flowing rivers. While freshwater hydrology became a critical component for tidal irrigation on the Cooper River, the Wando River's limited watershed did not generate enough flow to initiate this "hydraulic machine." Over millennia, the Wando's ebb and flow through the maritime floodplains created an interwoven chain of creeks and tributaries. These tributaries "arise from low, springy or marshy lands, and, as they branch out far and wide, innumerable navigable creeks are every way formed throughout the country." To utilize this environment, planters had to construct earthen barriers to prevent the brackish tidewater from flowing into these low-lying watercourses.[77]

Richard Beresford's use of tidal creeks reflected how planters utilized other environments besides small-stream floodplains to cultivate inland rice. Emigrating from Barbados to Charleston in 1683, Beresford began his career as a merchant owning a one-quarter share of the *Mary of Carolina*. He became active in politics, serving on the Grand Council, he represented Berkeley and Craven counties in five General Assemblies, and he was a member of the First Royal Assembly. Beresford's political ambitions correlated with his land acquisitions. Between 1690 and 1714, he received nine proprietary grants totaling 5,040 acres. Charleywood,

[76] Alexander Moore, "Daniel Axtell's Account Book and the Economy of Early South Carolina," *South Carolina Historical Magazine* 95 (October 1994): 299; Newington's rice field acreage was determined by averaging the annual amount of rice produced equaling 10,000 pounds, which is 222 bushels/annual @ 45 lbs/bushel. A 1712 report notes that a "good crop" between 30–40 bushels/acre was common for period rice plantations, so 222 bushels divided by 30 bushels/acre = 7.4 acres. Merrens, *Colonial South Carolina Scene*, 45.

[77] Quote: Chalmers, "A Sketch of the Climate," 316; Russell Austin Willis, "Genetic Stratigraphy and Geochronology of Last Interglacial Shorelines on the Central Coast of South Carolina," (M.S. thesis, Louisiana State University, 2006), 5–12; Lucy McCartan, E.M. Lemon, Jr., and R.E. Weems, *Geologic Map of the Area between Charleston and Orangeburg, South Carolina*, map I-1472 (Washington, D.C.: United States Geological Survey, 1984); Ralph H. Willoughby and W.R. Doar, III, *Solution to the Two-Talbot Problem of Maritime Pleistocene Terraces in South Carolina* (Columbia: South Carolina Department of Natural Resources, Geological Survey, 2006).

a plantation named after a Hertfordshire manor, derived from a series of seven 1711 grants of 4,350 acres.[78]

Building capital from the mercantile trade, cattle ranching, and naval stores, Beresford amassed a sizable labor force for rice cultivation. As one Santee Huguenot émigré explained, growing rice "can only be done at great expense and only rich people could undertake it." By 1715, Beresford had acquired fifty enslaved people. Seven years later, Beresford had doubled his enslaved population and was well within the 29 percent of Carolina slaveholders with thirty people or more by the 1720s. Beresford shuttled this substantial labor force between his seven plantations, where they herded cattle, extracted tar and pitch, and grew mulberry trees, corn, and rice. However, Beresford's labor force was not large enough to alter more than the seventy-five acres of the Charleywood rice fields. Compared to period rice plantations, more laborers were needed to build earthen barriers separating saline from fresh water zones, and to properly maintain the human-made environments against the natural tidal surges.[79]

Charleywood rice fields contained the same basic structure as at Pooshee, yet the subtle elevation change on the Wando River floodplains created a different aesthetic. Whereas Pooshee Swamp consisted of a relatively straight watercourse from the spring, Charleywood's tidal creeks came from multiple directions, wrapping around subtle highland knolls, and converging in Guerin Creek. Pooshee's rice field consisted of a single system of two dams bordering the rice. Charleywood, however, relied on dams to partition seventy-five acres into seven field divisions that, when compared to flooding a single unit, allowed improved irrigation control. As early inland fields were limited to narrow watercourses, their acreage did not compare to later tidal systems sprawling across broad floodplains. Beresford had to work around the low profile of the floodplains, where "highland" that enclosed the rice fields was only four feet above the rice fields.

[78] Agnes Leland Baldwin, *First Settlers of South Carolina, 1670–1700* (Easley, SC: The Southern Historical Press, 1985), 22; N. Louise Bailey and Walter Edgar, eds., *Biographical Directory of the South Carolina House of Representatives, Volume II: The Commons House of Assembly, 1692–1775* (Columbia: University of South Carolina Press, 1977), 77–78; Society for the Propagation of the Gospel in Foreign Parts, South Carolina, ser. A, vol. 16, 120; Colonial Land Grants, 1688–1799, SCDAH, vol. 39, 102–103, 311, 315.

[79] McClain and Ellefson, "Letter from Carolina," 392; Society for the Propagation of the Gospel in Foreign Parts, South Carolina, ser. A, vol. 16, 120; Coclanis, *Shadow of a Dream*, 98.

To retain water within the subtle elevation change, enslaved laborers had to construct embankments flanking each side of the field. Part of the challenge of establishing rice within this area was the limited elevation change separating the fields from the settlement. Compared to Pooshee's ten-foot and Newington's forty-foot difference between the fields and the settlements, Charleywood's four-foot difference did not allow much depth to the rice fields. Beresford had to build retaining embankments around the fields to hold more floodwater as the plants grew in size. Early inland cultivators had to pay attention to subtleties of the land and judge when an impounded field was too big to draw water effectively on and off the fields. By subdividing the fields, even in situations where water directly flowed from one field to the next, cultivators could manage the amount of water on individual plots and flood the entire crop more consistently in shorter distances with a low elevation run compared to one elongated field with a greater elevation change. Charleywood fields averaged five acres for five of the divisions, yet expanded to an average of twenty-five acres for the remaining two divisions. Even with planters modifying their fields, problems resulted from having to flood each division in order from the lowest elevation to the highest.[80]

With a general understanding of reservoir irrigated rice cultivation in a growing market economy, European colonists began shifting settlement patterns by the first decade of the eighteenth century toward low-lying small-stream floodplains and bottomlands. Incorporating technological and agricultural knowledge within new wetland boundaries, planters increased yields by approximately twenty bushels per acre before 1740. Rice cultivation expanded rapidly: in 1699, South Carolina's first major export was 300 barrels to England; by 1714, it was 11,000 barrels. Changing settlement patterns reveal the drive to produce this cash crop. A map drawn by John Thornton and Robert Morden in 1685 shows settlements concentrated around Charles Town and Goose Creek with limited populations noted in outlying areas, primarily lining navigable rivers. By 1711, the Edward Crisp map shows growth in Charles Town and Goose Creek plus additional plantations along the Ashley, Cooper,

80 "A Plan of Charleywood Plantation," March 1788, John McCrady Plat Collection, no. 954, Charleston County Register of Mesne Conveyance (CCRMC), Charleston, SC; Judith Carney explains that inland rice fields similar to Charleywood are reminiscent of West African mangrove systems in *Black Rice*, 86–88. Edda L. Fields-Black provides a more detailed discussion of Rio Nunez cultivators' specialization of mangrove rice environments in *Deep Roots*, 36–46, 57–106; for plantation elevations, see: USGS The National Map, http://nationalmap.gov/ (accessed April 9, 2019).

and South Edisto (Pon Pon) Rivers, along with inland holdings. Historian Converse Clowse estimated that the Proprietary government granted at least 200,000 acres between 1694 and 1705. About 100,000 acres of that land issued between 1698 and 1705 came from the headrights of enslaved Africans.[81]

The American rice economy, like the cultivation process, slowly took hold in South Carolina. English merchants initially discouraged South Carolina planters from competing on the global market because of low demand, high shipping prices, and inconsistencies in rice quality. Unlike Barbados sugar or Virginia tobacco, South Carolina rice did not tie into the English commission system. Instead, South Carolina factors originally sold rice on the domestic market centered in Charles Town. Planters could not get credit from English merchants, and rice planters had to seek out much-needed credit to expand their plantations by dealing directly with local merchants. Planters' connections with local merchants, either through family or social networks, created trade, illegally bypassing the English Crown: for example, the land-owning St. Julien and Ravenel families of Middle St. John's Berkeley worked together with the Le Serrurier and Mazyck families of Charles Town's burgeoning shipping and mercantile business. Isaac Mazyck aligned with his father-in-law, Jacques Le Serrurier, Sr., and Peter St. Julien went into business with his brother-in-law Jacques Le Serrurier, Jr. While other colonial cash crops (with the exception of tobacco) had to go through English networks, which slowed distribution and sales while increasing taxation, Charles Town rice factors could quickly sell the grain at the highest price throughout the world. This increasing wealth for both the factor and the planter produced capital that was reinvested into these plantation enterprises.[82]

[81] Norris, *Profitable Advice*, 40; Oldmixon, *The British Empire in America*, 519; Richard L. Haan, "The 'Trade Do's Not Flourish as Formerly': Ecological Origins of the Yamassee War of 1715," *Ethnohistory* 28 (Autumn 1981): 350–351; "A New Map of Carolina," John Thornton and Robert Morden (1685), University of North Carolina, North Carolina Map Collection, Chapel Hill, NC; "A Compleat [*sic*] Description of the Province of Carolina," Edward Crisp (1711) Geography and Map Division of the Library of Congress, Washington, D.C.; Clowse, *Economic Beginnings*, 131.

[82] R.C. Nash, "The Organization of Trade and Finance in the Atlantic Economy: Britain and South Carolina, 1670–1775," in *Money, Trade, and Power: The Evolution of Colonial South Carolina's Plantation Society*, eds. Jack P. Greene, Rosemary Brana-Shute, and Randy J. Sparks (Columbia: University of South Carolina Press, 2001), 77–81; Russell R. Menard, "Financing the Lowcountry Export Boom: Capital and Growth in Early South Carolina," *The William and Mary Quarterly*, 3rd ser. 51 (October 1994): 671–673; Van Ruymbeke, *From New Babylon to Eden*, 204.

By 1715, Carolina planters had settled as far south as the Edisto River. These plantations were located near the boundary of the Yamasee Indian lands. The ever-expanding colonists encroached onto Yamasee lands, with stray livestock foraging on vegetation competing with deer and other subsistence game. Frustrated, the tribe attacked settlers on April 15, 1715. The Yamasee War created two years of economic and agricultural stagnation and set in motion changes in political and economic structure that took more than fifteen years to overcome. The war devastated Carolina's southernmost plantations, destroying "near 400 of the [white] Inhabitants ... with many Houses and Slaves, and great numbers of Cattle." Yamasee destruction sent Carolina into an economic depression. Exports of salt meat declined by 2,413 barrels and rice by 4,438 in 1717, compared to 1712. Although meat exports continued at depressed numbers until 1731, annual rice production grew from 22,000 barrels in 1722 to 41,000 in 1730.[83]

Colonial expansion into the Carolina frontier stalled for fifteen years after the Yamasee War. Angered by the Proprietors' inability to handle the Native American attacks, colonists overthrew the proprietary government in 1719. With the removal of the provincial government and subsequent closing of the colonial land office, official transfer of land all but ceased. Only illegal settlement from squatting pushed European and African agricultural practices further into the South Carolina frontier. Carolina's economic recovery and colonial expansion began in 1730 when the English Crown bought out seven of the Lord Proprietors and claimed Carolina as a royal colony. This change in government officially ended the proprietary era in South Carolina and led to a shift in inland plantation structure and land distribution. With the Crown in charge, the royal government reopened the land office. The royal government distributed public lands liberally to prospective planters. The government renewed the Proprietors' headright policy, awarding fifty acres to each settler and fifty acres for each imported enslaved laborer. Royal authorities also permitted colonists to purchase lands at £20 sterling per 1,000 acres and a quitrent of half a pence per acre. This dramatic increase in grants spreading over new and uncultivated landscapes led to a new era of inland rice cultivation, where the topography

[83] Otto, *Southern Frontiers*, 37–38; Meriwether, *The Expansion of South Carolina*, 6; Stephen J. Oatis, "A Colonial Complex: South Carolina's Changing Era of the Yamassee War, 1680–1730," (Ph.D. diss., Emory University, 1999), 397–411.

challenged cultivators to incorporate new flooding and drainage techniques in their rice fields.[84]

From the founding of Carolina to the end of the proprietary period in 1729, colonists and slaves transformed rice from one of several experimental commercial ventures into the central cash crop fueling their economy. Rice cultivation techniques changed from the simple broadcasting of seeds on highland fields to intricately irrigated low-lying embanked divisions. For this change to happen, colonists had to alter their perception of wetlands. Europeans initially viewed wetlands in an unfavorable light, but colonists began associating the low-lying hardwood forests with optimism as they began to understand how to cultivate the crop in that environment. By the end of the proprietary period, colonists and slaves were cultivating rice in an array of microenvironments, from small-stream floodplains to hardwood bottomlands. With people beginning to understand the intricacies of the Carolina landscape by the third decade of the eighteenth century, planters would begin using this knowledge of cultivating rice on a grander scale in the second half of the colonial era.

[84] Otto, *Southern Frontiers*, 38; Oatis, "Colonial Complex," 397–411; Catherine Armstrong, *Landscape and Identity in North America's Southern Colonies from 1660 to 1745* (London: Ashgate, 2013), 168.

3

The "Golden Mines of Carolina"

Expansion of the Inland Complex, 1730–1783

From 1730 to the eve of the Revolution, planters expanded inland rice cultivation from the small stream floodplains to broad bottomland basins. As market demand for rice and the enslaved labor population increased, rice cultivation nestled its way into a variety of watersheds. The move from simple stream floodplains to intricate bottomland basins challenged those planters attempting to expand their rice fields. Changes in the landscape required a holistic understanding of water control combined with larger labor forces necessary to carry out those visions. Planters balanced cultivation output with the realities of environmental limitations, while they increasingly managed enslaved communities to alter even more terrain. Yet during the final two decades of the colonial era, inland rice planters began seeing declining yields brought about by human-induced "natural" disasters, from floods to declining soil fertility.

This emerging rice landscape was created by the developing Atlantic world market economy, increasing population in the Carolina frontier, and Europeans' and Africans' technological transfer of cultivation practices to the colony. Interwoven through these themes were successive shifts in the perception of land. How people saw the colony was reflected in their settlement patterns and attention to healthy environments. The Atlantic world market economy encouraged planters to commit further to the capital-intensive grain. Expanding free and enslaved populations, coupled with the reopening of the colonial land grant office, placed more people onto the Carolina frontier. The "peopling" of frontier environments presented colonists and slaves with new choices of how to develop a variety of ecosystems. Finally, a greater understanding of rice culture enabled planters to cultivate undeveloped wetlands. Perhaps the greatest development in inland rice cultivation during this period was the introduction of new water management techniques. Inspired by the advent of tidal irrigation, inland rice planters initiated more capital-intensive internal improvements to move water through the low-lying floodplains.[1]

The developing market economy in the Atlantic world led to increasing export of South Carolina rice. Demand for the grain had exploded in both Europe and America by 1730, which initiated more capital investment in rice plantations. One key pull factor led to this demand for South Carolina rice. The British Parliament relaxed the Navigation Acts for rice in 1731. This change in policy allowed for the direct export of rice on British ships headed to Southern Europe, which consumed 20 percent of colonial rice exports by the 1730s. In 1699, Charles Town merchants exported 291 barrels of rice; by 1715, the total rose to 5,262 barrels. However, fifteen years later, South Carolina merchants exported 44,385 barrels and by 1745 the total had grown to 63,433 barrels. Despite rising rice exports, prices fluctuated dramatically during a twenty-one year period beginning in 1739. A series of economic depressions in England and Charleston – stemming from King George's War (1739–1748), the Stono Rebellion (1739), and yellow fever (1739, 1745, 1748, 1758) and smallpox (1738, 1758, 1760) epidemics – led to deflated rice prices. Nonetheless, shipping increased between Europe and North America after the end of King George's War in 1748, and lower transportation costs, the consequence of faster shipping and higher volume, increased profits. English merchants could purchase less expensive rice from India or Africa at the end of the seventeenth century, yet they obtained a higher quality grain at a lower price from South Carolina by the mid-eighteenth century.[2]

The South Carolina rice market began another upward economic and manufacturing cycle in 1760. International events advanced rice prices in the second half of the eighteenth century, spurring an increase in

[1] "Peopling" is derived from Bernard Bailyn, *The Peopling of British North America: An Introduction* (New York: Alfred A. Knopf, Inc., 1986).

[2] Henry C. Dethloff, "The Colonial Rice Trade," *Agricultural History* 56 (January 1982): 236; Stephen G. Hardy, "Colonial South Carolina's Rice Industry and the Atlantic Economy: Patterns of Trade, Shipping, and Growth, 1715–1775," in *Money, Trade, and Power: The Evolution of Colonial South Carolina's Plantation Society*, ed. Jack P. Greene, Rosemary Brana-Shute, and Randy J. Sparks (Columbia: University of South Carolina Press, 2001), 116; John J. McCusker, "Table Eg1160–1165, Rice exported from South Carolina and Georgia: 1698–1790," in *Historical Statistics of the United States*, Millennial Edition Online, ed. Susan B. Carter et al. (New York: Cambridge University Press, 2006), 5-763, 5-764.

Lowcountry rice production. European demand for rice grew dramatically after a series of poor English and European grain harvests during the late 1760s and early 1770s. This motivated the British Parliament to remove tariffs and import more of the South Carolina cash crop. Climate fluctuations between 1767 and 1771 created abnormally high rainfall in the summer and autumn, while cold and long winters prolonged snow cover. The cumulative moist seasons created damaging effects on European grain production, with wet autumns diminishing wheat production in the European lowlands and long winters depleting hay crops in the uplands. These environmental hardships suffered by European grain producers, and the relaxation of English trade restrictions, led to a 50 percent increase in imports of Carolina rice between ca.1760 and ca.1775.[3]

Scholars have argued that this dramatic increase in South Carolina rice exports was a direct reflection of the development of tidal irrigation. While planters' increasing implementation of this cultivation technique is well documented, a central problem persists in this interpretation. Historians have interpreted the development of South Carolina rice cultivation in a teleological fashion, in which a series of technological innovations – first through inland and then through tidal cultivation – led to a more efficient method of growing rice. According to this historiography, tidal rice culture was the highest form found in South Carolina.[4] Two schools of thought disagree about how this agriculture evolved. The first school cites that tidal rice "was a distinct, individual innovation" apart from the earlier inland rice culture. These scholars argue that the two cultivation methods were different in technological understanding and

[3] Peter A. Coclanis, "Rice Prices in the 1720s and the Evolution of the South Carolina Economy," *Journal of Southern History* 48 (November 1982): 539–543; R.C. Nash, "South Carolina and the Atlantic Economy in the Late Seventeenth and Eighteenth Centuries," *Economic History Review*, New Series 45 (November 1992): 679, 684, 686–689, 692; Hardy, "Colonial South Carolina's Rice Industry and Atlantic Economy," 116. For eighteenth-century climate fluctuations and European agriculture, see: Christian Pfister, "Climate and Economy in Eighteenth-Century Switzerland," *Journal of Interdisciplinary History* 9 (Autumn 1978): 233–234, 239–240.

[4] David L. Coon, "The Development of Market Agriculture in South Carolina, 1670–1785," (Ph.D. diss., University of Illinois at Urbana-Champaign, 1972), 182–184; Russell R. Menard, "Economic and Social Development of the South," in *The Cambridge Economic History of the United States*, vol. I: *The Colonial Era*, ed. Stanley L. Engerman and Robert E. Gallman (Cambridge: Cambridge University Press, 1996), 282–283; Peter A. Coclanis, *The Shadow of a Dream: Economic Life and Death in the South Carolina Low Country, 1670–1920* (New York: Oxford University Press, 1989), 66–68, 96–97; Henry C. Dethloff, *A History of the American Rice Industry, 1685–1985* (College Station: Texas A&M University Press, 1988), 70–72.

topographical placement. A second school came to argue that inland cultivation served as a springboard for tidal development. These interpretations explain that inland production provided an economic foundation and general technological understanding for planters to gradually evolve their irrigation methods along the tidal floodplains. While this second school is correct in that inland cultivation did provide a foundation for tidal planting, their interpretation simplifies the transition between the two technologies. Scholars believe Lowcountry planters slowly abandoned their static inland fields for the ever-evolving tidal agroecology, but they did not. Instead, inland production coexisted with tidal production, and it too developed increasingly complex forms.[5]

Shifts in internal improvements, labor control, and economies of scale associated with the introduction of tidal irrigation by 1738 led to changes in inland cultivation and the landscapes they utilized. Inland rice culture was not static in colonial South Carolina. Inland planters, like their tidal counterparts, aggressively sought out new landscapes to improve their agricultural conditions and utilized water control methods that developed in unison with those of the emerging tidal planters. Rice planters, in general, were peers, related by marriage or blood, and of equal social, economic, and political stature. These circles exchanged agricultural ideas

[5] S. Max Edelson, "Planting the Lowcountry: Agricultural Enterprise and Economic Experience in the Lower South, 1695–1785," (Ph.D. diss., Johns Hopkins University, 1998), 216. Scholarship documenting tidal cultivation as a distinct innovation separate from inland cultivation is found in Ulrich Bonnell Phillips, *Life and Labor in the Old South* (Boston: Little, Brown, and Company, 1929), 116; Lewis Cecil Gray, *History of Agriculture in the Southern United States to 1860*, 2 vols. (Washington, D.C.: Carnegie Institution of Washington, 1933; reprint, Gloucester, MA: Peter Smith, 1958), I: 279–280; Dethloff, "Colonial Rice Trade," 238; James M. Clifton, "The Rice Industry in Colonial South Carolina," *Agricultural History* 55 (July 1981): 275; Julia F. Smith, *Slavery and Rice Culture in Low Country Georgia, 1750–1860* (Knoxville: University of Tennessee Press, 1985), 21; Dethloff, *Rice Industry*, 70–71. For scholarship documenting the evolution of tidal irrigation, see Sam B. Hilliard, "Antebellum Tidewater Rice Culture in South Carolina and Georgia," in *European Settlement and Development in North America: Essays on Geographical Change in Honour and Memory of Andrew Hill Clark*, ed. James R. Gibson (Toronto: University of Toronto Press, 1978), 97; Coon, "Development of Market Agriculture," 182–183; Robert M. Weir, *Colonial South Carolina: A History* (New York: KTO Press, 1983; reprint, Columbia: University of South Carolina Press, 1997), 150; Joyce E. Chaplin, *An Anxious Pursuit: Agricultural Innovation and Modernity in the Lower South, 1730–1815* (Chapel Hill: University of North Carolina Press, 1993), 237, 242; Edelson, "Planting the Lowcountry," 216. For critique of teleological rice history, see Francesca Bray, *The Rice Economies: Technology and Development in Asian Societies* (Oxford, UK: Basil Blackwell, 1989; reprint, Berkeley: University of California Press, 1994).

as freely as they exchanged current events. In essence, inland and tidal cultivation in the colonial Lowcountry share a similar story. Each method advanced in the developing frontier landscape. To implement these more sophisticated hydrological systems, both inland and tidal planters relied upon a growing enslaved labor population to create precisely crafted embankments, canals, and drains. While agricultural knowledge transferred between planters, inland planters also relied upon and contributed to advancements promoted along the tidal rivers.

The peopling of the expanding Carolina frontier accounted for increasing rice output and coincided with the shifting rice market. In 1729, the Crown's purchase of the proprietors' rights signaled a new era of expansion and land accumulation. More than a decade had passed since the proprietors closed the land office. During that time, colonists had to acquire land through shifty means. Individuals either purchased land through the proprietors in England or placed tentative claims domestically through "illegal" surveys. However, the reopening of the land office, a brief stability in rice markets during the 1720s, removal of some export tariffs, and new bounties placed on naval stores fueled a land boom in the 1730s. As colonists pushed further out onto the frontier – approximately forty miles from Charleston, according to S. Max Edelson – the newly appointed Governor Robert Johnson issued a "township scheme" and fortification plan in 1730. Townships attracting an influx of immigrants of Scottish, Swiss, and German descent, combined with fortifications along the outlying colonial boundaries, provided a line of defense against the Native Americans, French, and Spanish.[6]

Colonists' demand for land during the first decade of the royal period generated sustained speculation and acquisitions. The Middletons, Izards, Cattels, and Balls capitalized on rice cultivation during the first two decades of the eighteenth century and could afford large tracts of land on the reopened Carolina frontier. Their purchase of land further away from Charles Town represented the speculative spirit as these entrepreneurs did not know the topographic details within their undeveloped properties, only that their newly acquired land possessed the possibility for new rice plantations. Edelson explained how a division in settlement patterns existed between the "core," "secondary," and "frontier" zones. He defined the core zone as an egg shaped boundary encompassing the

[6] Weir, *Colonial South Carolina*, 111–117; S. Max Edelson, *Plantation Enterprise in Colonial South Carolina* (Cambridge, MA: Harvard University Press, 2006), 127, 129–130.

watersheds of the Stono, Ashley, Cooper, and Wando Rivers. With Charles Town as the center, the core zone of settlement followed the four rivers into the interior. A secondary zone formed a crescent between the Edisto and the Santee Rivers, while the frontier zone extended one hundred miles up and down the coast and fifty miles inland from Charles Town.[7]

During the mid-eighteenth century expansion, the secondary zone offered new watersheds for rice cultivation and blank slates for water control. Larger landholdings in the secondary zone presented more opportunities for rice cultivation, as the average size of a plantation within the core zone was 266 acres while the average size in the secondary and frontier zones was 500 acres. Edelson suggests that just under one-half of the land was suitable for growing rice in the secondary zone, compared to approximately one-third of the land in the core zone. He explains that lands close to Charleston did not possess the broad wetlands that characterized larger tracts in the frontier. According to Edelson's calculations, the frontier tracts averaged 372 acres of wetlands suitable for rice cultivation compared to core zone tracts that averaged 204 acres for rice cultivation.[8]

Coinciding with the expanding plantation lands during the 1730s was the increasing importation of enslaved Africans. The black population grew by 19,155 people, or 95 percent between 1730 and 1740. Although South Carolina had a "black majority" by 1710, the population of Africans peaked at 90 percent of the total Lowcountry population in 1740.[9] As slavery expanded during the 1730s, the newly available work forces enabled planters to solve problems of labor shortages on the inland rice fields. Rice was a labor-intensive commodity, and the availability of labor determined how much rice the planters could grow. Limited labor contributed to limited cultivation. However, massive slave importation slowed as a result of the 1739 Stono Rebellion. In an effort to prevent

[7] Weir, *Colonial South Carolina*, 113–114; David B. Ryden and Russell R. Menard, "South Carolina's Colonial Land Market: An Analysis of Rural Property Sales, 1720–1775," *Social Science History* 29 (Winter 2005): 605; Edelson, *Plantation Enterprise*, 129–141, Table A.4: "Geography of lowcountry settlement zones," 275.

[8] Edelson, *Plantation Enterprise*, 138, 140, Table A.6: "Settled and undeveloped lands, 1732–1775," 280; Table A.9: "Size, dispersion, and development of land, 1732–1775," 281–282; Table A.10: "Capacity of land for rice agriculture, 1732–1775," 282.

[9] Coclanis, *Shadow of a Dream*, 67–68, 64, Table 3-1: "South Carolina population estimates"; Carville Earle, *The American Way: A Geographical History of Crisis and Recovery* (Lanham, MD: Rowman and Littlefield, 2003), 283–284.

future slave rebellions, the South Carolina House of Commons passed laws to control the size of African populations on Lowcountry soil. The Negro Act of 1740 limited the numbers of incoming Africans for most of the decade – specifically banning slave importation between 1741 and 1743 – and attempted to immobilize enslaved African-Americans' freedoms until the end of the antebellum period. It curbed slaves' ability to travel, assemble in groups, raise food, earn money for personal use, and receive an education. While historians debate the effectiveness of this Act, where a consensus is formed is that "a degree of compromise and fluidity between blacks and whites" came into existence. However, in the 1740s, despite subtle agency, enslaved African Americans began an increasingly repressive chapter in the history of inland rice cultivation.[10]

The answer to the planters' labor problem came from the gradual domestic extension of credit for Lowcountry plantations. While land became readily available to rice planters after 1730, the lack of capital to purchase land and labor suppressed potential expansion into the frontier zone. Unlike their counterparts in the British West Indies and Virginia, South Carolina planters did not receive capital advancements directly from British merchants. Instead, people who could not pay up front had to finance land, slaves, and manufactured goods from domestic merchants. This combination of available land, labor, and capital led opportunistic planters to develop larger amounts of acreage. However, the fluctuation of rice prices, the Stono Rebellion, and disease placed temporary roadblocks in front of the ever-expanding risiculture. The ten-year period from 1730 to 1740 saw peaks and troughs in rice prices, slave importation, and land improvement. However, South Carolina's annual average export of rice jumped from 44,385 barrels in 1730 to

[10] Coclanis, *Shadow of a Dream*, 57–58, 64; Edelson, *Plantation Enterprise*, 64; Weir, *Colonial South Carolina*, 194. Russell R. Menard, "Slavery, Economic Growth, and Revolutionary Ideology in the South Carolina Lowcountry, in *The Economy of Early North America: The Revolutionary Period, 1763–1790*, ed. Ronald Hoffman et al. (Charlottesville: University Press of Virginia, 1988), 259; Ira Berlin, "Time, Space, and the Evolution of Afro-American Society on British Mainland North America," *American Historical Review* 85 (February 1980): 58–61; Peter H. Wood, *Black Majority: Negroes in Colonial South Carolina from 1670 through the Stono Rebellion* (New York: W.W. Norton & Co., 1974), 323–325; Jack Shuler, *Calling Out Liberty: The Stono Slave Rebellion and the Universal Struggle for Human Rights* (Oxford: University of Mississippi Press, 2009), 99, 101–102, quote: 104.

96,926 barrels in 1740. By 1773, the combination of inland and tidal cultivation contributed to the peak colonial export of 164,704 barrels.[11]

A critical component of the increase in rice output during the colonial period was the advent of tidal irrigation. Tidal cultivation, although it occurred close to inland fields, relied on different topographical and hydrological conditions. Tidal rice planters irrigated the fields with the fresh-water river tides. Planters used the rising tide, or "flow," to irrigate the rice and then the falling tide, or "ebb," to drain the fields. Permanent embankments and surrounding interior ditches kept high water out of fields and floodwater in, and they allowed proper draining of fields. As with inland fields, hand made "rice trunks" controlled water flowing in and out of the embankments, yet tidal rice farmers modified these wooden devices to allow a multidirectional flow of water. To achieve this water control, planters designed gates to cover both ends of the trunk. When flooding fields, slaves raised the exterior gate closest to the incoming river, while the interior gate pivoted on a hinge as water flowed through it and into fields from the force of the tide. Once the tide changed direction, slaves closed both gates to prevent impounded water from leaving the fields. After the desired time elapsed for irrigating the fields, slaves raised the interior gate to allow water to flow out of the pivoted exterior gate, preventing resurgent tidal waters from flowing into the fields. Tidal rice fields were sub-divided into smaller plots to efficiently control water flow, and these subdivisions were called "quarter" divisions because they were originally a quarter of an acre. Planters connected these divisions to a network of canals, ditches, and drains to irrigate the crop and power rice mills.[12]

[11] R.C. Nash, "The Organization of Trade and Finance in the Atlantic Economy: Britain and South Carolina, 1670–1775," in *Money, Trade, and Power: The Evolution of Colonial South Carolina's Plantation Society*, ed. Jack P. Greene, Rosemary Brana-Shute, and Randy J. Sparks (Columbia: University of South Carolina Press, 2001), 93–94; Coclanis, *Shadow of a Dream*, 65–66, 82; McCusker, "Table Eg1160–1165, Rice exported from South Carolina and Georgia," 5-763, 5-764.

[12] Sam B. Hilliard, "The Tidewater Rice Plantation: An Ingenious Adaptation to Nature," in *Geoscience and Man* 12, ed. H.J. Walker (Baton Rouge: Louisiana State Press, 1975), 57–66; Mart A. Stewart, *"What Nature Suffers to Groe": Life, Labor, and Landscape on the Georgia Coast, 1680–1920* (Athens: University of Georgia Press, 1996), 98–116. Tidal cultivation is further discussed in Judith Carney, *Black Rice: The African Origins of Rice Cultivation in the Americas* (Cambridge, MA: Harvard University Press, 2001); Philip D. Morgan, *Slave Counterpoint: Black Culture in the Eighteenth-Century Chesapeake and Lowcountry* (Chapel Hill: University of North Carolina Press, 1998); William Dusinberre, *Them Dark Days: Slavery in the Inland Rice Swamps* (New York: Oxford University Press, 1996); Chaplin, *Anxious Pursuit*. Earlier works include: Duncan Clinch

When acknowledging the dramatic increase in labor devoted to rice and the differences in rice export trends during the colonial period, one has to realize a major influence came from the new irrigation technology. Tidal cultivation accessed consistent water sources from the adjoining rivers and used systematic flooding schedules to control competing weeds. However, the new technological implementation went beyond the simple shift in waterpower. Tidal planters' drive to amass a large enough labor force to move the tremendous amount of earth required for irrigation canals, drains, and ditches also influenced inland planters' visions for their own holdings.

By shaping inland floodplains in a similar design to the tidal-influenced riverbanks, with increasing numbers of enslaved laborers to perform Herculean tasks, inland planters incorporated new lessons of irrigation and labor control into their own fields. By 1740, inland planters and their enslaved laborers were using new modes of water control in the core and secondary zones. The developing irrigation methods required rice cultivators to take "command of water" to secure systematic flooding and draining of fields. Inland planters also sought solutions to relieve pressure from freshets breaching reservoir dams. Flanking canals, which were dredged waterways that abutted exterior field embankments, provided a solution to this problem by diverting water around the impounded rice fields. Planters referred to flanking canals as "wasteway" or "washway" drains. The canals stretched the length of the field system. As naturalist William Bartram noted, these reservoirs were sometimes connected to "sluices to let ye redundant waters out." A similar concept of channeling water and relieving pressure from milldams also existed. Millowners installed canals that redirected water downstream from the reservoirs to prevent freshets from breaching their dams. Although millowners did not use the water released into their wasteway drains, Lowcountry inland rice

Heyward, *Seed from Madagascar* (Chapel Hill: University of North Carolina Press, 1937; reprint, Columbia: University of South Carolina, 1993); J.H. Easterby, ed., *The South Carolina Rice Plantation, as Revealed in the Papers of Robert F. W. Allston* (Chicago: University of Chicago Press, 1945); Elizabeth W. Allston Pringle, *Chronicles of Chicora Wood* (New York: Charles Scribner's Sons, 1922); Arney R. Childs, ed. *Rice Planter and Sportsman: The Recollections of J. Motte Alston, 1821–1909* (Columbia: University of South Carolina Press, 1953); David Doar, *Rice and Rice Planting in the South Carolina Lowcountry* (Charleston: Charleston Museum, 1936; reprint, 1970); James M. Clifton, ed., *Life and Labor on Argyle Island: Letters and Documents of a Savannah River Rice Plantation, 1833–1867* (Savannah, GA: Beehive Press, 1978); Phillips, *Life and Labor*; David Duncan Wallace, *The History of South Carolina*, 4 vols. (New York: American Historical Society, 1934); Gray, *Agriculture in the Southern United States.*

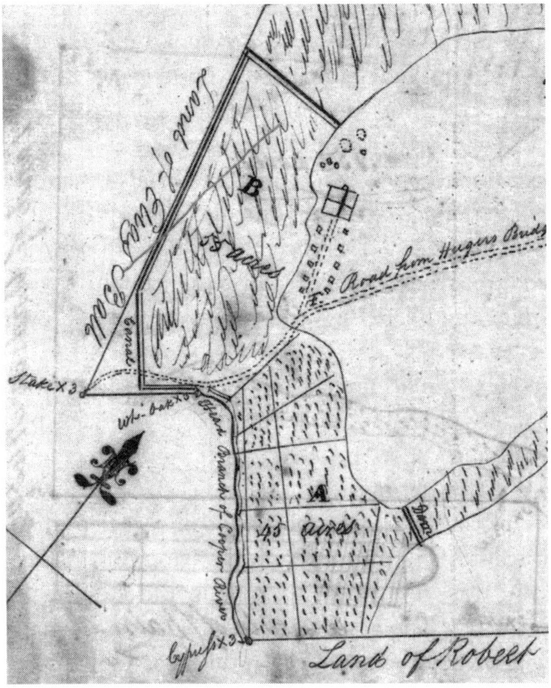

FIGURE 3.1 Example of a flanking canal, marked with a dark line around division "B," in relation to the lower rice fields of division "A." Water flows in this illustration from top to bottom.
"Plan of Windsor Plantation, March 1790" (detail), Book D7: 199, Charleston County Register Mesne Conveyance, Charleston, SC

planters channeled their wasteway water from the dam to the fields. Rice cultivators accomplished this water redirection by inserting trunks between the flanking canal and the upper and lower portion of each field division. Water would enter the field through the upper trunk and drain out of the field through the lower trunk. This adaptation allowed trunk minders to irrigate fields without having to flow each division simultaneously. The flanking canal also served as a gutter to capture any downward flowing water from higher ground perpendicular to the fields.[13]

[13] The term "diversion canals" is also used in describing flanking canals. Edelson, *Plantation Enterprise*, 103–109; Heyward, *Seed from Madagascar*, 13; Alexander Moore, "Daniel Axtell's Account Book and the Economy of Early South Carolina," *South Carolina Historical Magazine* 95 (October 1994): 292–296; Bartram quote: Richard I. Groening, Jr., "The Rice Landscape in South Carolina: Valuation, Technology, and Historical Periodization" (M.A. thesis, University of South Carolina, 1998), 79;

Flexibility to control water was essential when rice fields were on different cultivation schedules. Trunk minders could add or remove water as they saw fit without having to disrupt flood stages on adjoining fields. The staggering of flood schedules avoided possible depletion of impounded water, as springs and creeks could recharge the reservoirs before the next flood cycle. "Every planter has his reservoirs or ponds of water which are attended by drains and ditches that he can at any time set his plantation afloat," cited eighteenth-century attorney Timothy Ford, who observed "[the planter] must know more from his own judgment and observation than anything else, when, how often, and how long his fields must be under water."[14]

During freshets, trunk minders could release excess water through the flanking canals, bypassing the rice fields and relieving pressure on the back dam. Flanking canals provided partial relief to the inland planter who had too much water in the winter and early spring. Mathurin Gibbs, an inland rice planter in St. Stephen's Parish revealed this frustration in his nineteenth-century agricultural journal. On trying to keep water out of the fields as he was attempting to contain the natural resource in his reservoir, he wrote, "The season has been backwards and the floods of the rain have drowned the greater portion of rice sown in two fields and preventing my sowing the other fields till now." Ten days later, Gibbs explained, "The labor of man is vain for no sooner does his industry and perseverance remove the water from one field than it is filled by the rain and the labors … applied to be again disappointed." The ability to control water, harnessing this natural resource when needed, became the primary concern of the inland planter. Inconsistent water flow would ultimately detour inland cultivation, yet the eighteenth-century use of flanking canals provided a new flexibility for rice cultivators attempting to control natural conditions.[15]

Planters who grew rice within narrow floodplains had to make different decisions compared to inland planters cultivating along the broad tidal floodplains. These separate microenvironments led to planters using new methods of drawing water onto and off of the rice fields. Windsor

Commissioners of Fortified Estates, "Plan for Resurvey of 3,038 Acres on Wambaw Swamp, St. James Parish, Surveyed by William Evans, Formerly Property of Elias Ball," ca.1786, South Carolina Department of Archives and History (SCDAH), Columbia, SC.

[14] Joseph W. Barnwell, ed., "Diary of Timothy Ford, 1785–1786," *South Carolina Historical and Geological Magazine* 13 (October 1912): 183.

[15] 1 June 1846, 11 June 1846, Mathurin Guerin Gibbs Plantation Register, South Carolina Historical Society (SCHS), Charleston, SC.

Plantation on the headwaters of the East Branch of the Cooper River, for instance, demonstrated how flanking canals took shape. Windsor's fields fit within the tight boundary of the Nicholson Creek floodplain. The elevation difference between pineland communities and the cypress hardwood forest varied between thirty to forty feet. The watershed was dramatic in elevation change compared to the five-to-ten-foot decline in elevation located five miles away on the Cooper River tidal floodplains. Through the eighteenth century, the Roche family optimistically surveyed four divisions within the confines of the scarp to the northwest and the Ten Mile Hill Formation to the southeast. The Roches relied on the predominant knoll forming Nicholson Creek's southern boundary to contain the inland rice fields. Forming a crescent shape around a forty-foot bluff, Nicholson Creek connects with Turkey Creek to form Huger Creek and serves as the headwaters of the Eastern Branch of the Cooper River. This bluff served as an optimal site for the Windsor house, slave settlement, and outbuildings.[16]

Planters placed their rice fields in relation to the low-lying topography, which dictated the slaves' positioning of the flanking canal. At Windsor, Patrick Roche ordered twelve enslaved laborers to sculpt fields out of the Nicholson Creek cypress bottomlands by 1725. Fishbrook Field, named after the neighboring plantation, was the result of cutting trees, removing cypress stumps, and shaping forty-five acres of land. Nicholson Creek's meandering channel passed the western boundary of Fishbrook Field, separated by an earthen embankment; Roche's slaves altered the natural watercourse by embanking a fifty-five-acre division and channeling the water toward a flanking canal. An upstream field division impeded the natural watercourse with an earthen dam and then redirected the creek around the western perimeter. A variation on this system consisted of two canals, flanking the fields on each side. Dual canals increased the

[16] John B. Irving, "Record of Windsor and Kensington Plantations, 1840–1888," Charleston Library Society (CLS), Charleston, SC; W.R. Doar, III, and Ralph H. Willoughby, *Revision of the Pleistocene Dorchester and Summerville Scarps, the Inland Limits of the Penholoway Terrace, Central South Carolina* (Columbia: South Carolina Department of Natural Resources, Geological Survey, 2006); Robert E. Weems and Earl M. Lemon, Jr., *Geology of the Bethera, Cordesville, Huger, and Kittredge Quadrangles, Berkeley County, South Carolina* (Washington, D.C.: United States Geological Survey, 1989); Donald J. Colquhoun, *Terrace Sediment Complexes in Central South Carolina*, (Columbia: University of South Carolina, 1965), 29–31, quote: 29; "Inventory of Ebenezer Roche, 3 July 1784" South Carolina Estate Inventories, A: 378, SCDAH; "For Sale, On Tuesday the 6th of July Next, On the Premises, All the Estate of Ebenezer Roch, Deceased," *South Carolina Gazette* (Charleston, SC), 14 June 1784: 1.

efficiency of moving water around fields during freshets and also provided additional flexibility in flooding and draining individual divisions. Flanking canals varied in length and width, relative to the size of the plantation watershed and rice fields. For example, John Coming Ball's Back River Plantation utilized a fifty-foot-wide northern flanking canal as its reservoir while the southern flanking canal drained impounded water off the fields.[17]

As the midcentury Lowcountry plantation enterprise became firmly entrenched within Atlantic markets, inland planters began to initiate more aggressive cultivation practices. For example, Gabriel Manigault started to transform the massive 12,000-acre Awendaw Barony in 1739 into an intricate grid of rice fields bordering the confluence of the Awendaw and Steed Creeks. By the end of the colonial period, Manigault's slaves constructed two large divisions with a total of 400 acres. Nestled in this confluence sat the reservoir-fed rice fields, separated from the brackish tides by large earthen embankments. Expansion of inland landscapes stemmed from increased enslaved labor, acquired agricultural knowledge, improved canal networks, and lands suitable for cultivation. The developing inland field systems took on a new aesthetic, moving away from small acreage within natural boundaries and toward larger field divisions with geometrically rigid embankments. In the case of Manigault, his choice of locating fields irrigated by fresh water reservoirs along tidal riverbanks close to the ocean secured more acreage compared to his older fields on small-stream floodplains located at Silk Hope Plantation.[18]

On the Wando River, Richard Beresford, Jr., forced upwards of 253 enslaved laborers to expand his father's inland rice fields. The younger Beresford inherited Charleywood and other sizable holdings upon his father's death from a falling tree limb in 1722. Richard, however, was only two years old at the time. Lieutenant Governor Thomas Broughton, executor of the Beresford estate, managed the plantation affairs for 10 percent of the annual profits until the younger Beresford reached twenty-one years of age. Educated in England and working as a

[17] "Plan of Windsor Plantation, March 1790," Deed Book D7: 199, Charleston County Register Mesne Conveyance (CCRMC), Charleston, SC; http://nationalmap.gov/; Charles Hateley to John Coming Ball, 6 August 1762, Ball Family Papers, South Caroliniana Library, University of South Carolina (USC), Columbia, SC.

[18] Maurice Alfred Crouse, "The Manigault Family of South Carolina, 1683–1783 (Ph.D. diss., Northwestern University, 1964), 350–352; Deed Book T: 248–259, A3: 99–107, CCRMC; Henry A.M. Smith, "The Baronies of South Carolina: Seewee Barony," *South Carolina Historical and Genealogical Magazine* 12 (July 1911): 115–116.

London merchant until the age of twenty-seven, Richard returned to Charles Town in 1747 to manage his plantation enterprise. Like his father, Beresford maintained an active life, representing the Parish of St. Thomas and St. Denis in three Royal Assemblies.[19]

During Beresford's occupation, Charleywood reached a total of 3,715 acres "at which fourteen hundred acres is exceedingly fine Rice-Land." Although not all of this advertised acreage was put under bank, Charleywood cultivators relied on variable creeks and sloughs to adequately capture water for inland rice irrigation. Beresford impounded water from tributaries flowing south from the Suffolk Scarp toward Guerin Creek and also from an extensive canal system flowing west from Fairlawn. Before Richard died in 1772, he included in his will that all property was to be sold and equally divided in eight shares. The Beresford executors divided the plantation into smaller tracts to entice more bidders. Merchant Peter Taylor of Whitehaven, England purchased the central 1,080-acre "Charleywood" tract plus the 300-acre "Well's" tract adjoining to the north for £24,000. S.C. Taylor's attorney, Thomas Smith, served as property manager for the absentee businessman and oversaw general plantation operations. Smith and Matthias Rash, Charleywood's overseer, coordinated the agricultural schedule until Taylor sold the property to Charles Cotesworth Pinckney and Edward Rutledge in 1788.[20]

Charleywood's rice fields encompassed nearly 600 acres of the Wando River floodplain. Whereas the banks resembled the broad and expansive floodplains of sought-after tidal rice landscapes, the Wando's brackish water made this floodplain useless to tidal planters. Inland planters channeled upland fresh water down to the brackish floodplains to grow rice. In comparison to the narrow Nicholson Creek watercourse, Charleywood's

[19] Society for the Propagation of the Gospel in Foreign Parts, South Carolina, Series A, vol. 16, 115; N. Louise Bailey and Walter Edgar, eds., *Biographical Directory of the South Carolina House of Representatives, Volume II: The Commons House of Assembly, 1692–1775* (Columbia: University of South Carolina Press, 1977), 78–79.

[20] Harriette Kershaw Leiding, *Historic Houses of South Carolina* (Philadelphia: J.B. Lippincott, 1921), 186–187; "Richard Beresford Will," Miscellaneous Inventories and Wills, Charleston County, 14:300, SCDAH; "The Following Lands, Belonging to the Estate of Richard Beresford, Esq., Deceased,"*South Carolina Gazette* (Charleston, SC), 27 October 1772: 3; Deed Book E4: 404–412, CCRMC; Matthias Rash, Charley Wood, to Peter Taylor, Whitehaven, Cumberland, 18 March 1773, Thomas Smith, Charles Town, to Peter Taylor, Whitehaven, 14 July 1773, Peter Taylor, New York, to Isabella Fleming Taylor, 16 January 1783, "A Return of Peter Taylor's Estate" [ca.1785], Peter Taylor to "My Dear Friend," 26 November 1788, Taylor Family Papers, USC; Deed Book B9: 254, CCRMC.

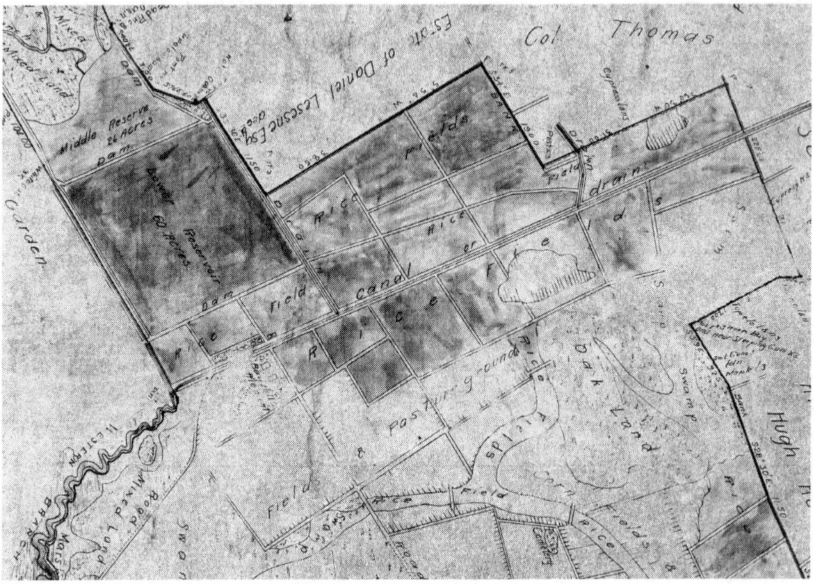

FIGURE 3.2 Charleywood rice fields.
"A Plan of Charleywood Plantation," March 1788, (detail) John McCrady Plat Collection, 2649, Charleston County Register Mesne Conveyance, Charleston, SC

topography provided a vast foundation for Beresford's massive enslaved labor force to carve an intricate grid-like formation of canals, ditches, embankments, and dikes. The new, expanded field system was built on Pleistocene deposits of clay and shell, which provided more effective water retention and higher nutrient yields, compared to the elder Beresford's earlier field system, which was a sandy-loam. To irrigate Charleywood's larger rice fields, enslaved cultivators relied on two reservoirs located on the Suffolk Scarp. The reservoirs impounded more than forty acres of water flowing from the meandering creeks and bays common to the Lower Coastal Plain scarps, while canals channeled the water in a linear downward motion.[21]

[21] "Richard Beresford Inventory, 23 January 1773," Miscellaneous Inventories and Wills, Charleston County, 94B: 423, SCDAH; Richard D. Porcher and Douglas A. Rayner, *A Guide to the Wildflowers of South Carolina* (Columbia: University of South Carolina Press, 2001), 40–41; "A Plan of Charleywood Plantation," March 1788, John McCrady Plat Collection, no. 954, CCRMC; Lucy McCartan, E.M. Lemon, Jr., and R.E. Weems, *Geologic Map of the Area between Charleston and Orangeburg, South Carolina* Map I-1472 (Washington, D.C.: United States Geological Survey, 1984); Mart Stewart discusses

Charleywood's settlement patterns shifted in relation to the plantation's rice cultivation. Early Charleywood inhabitants lived on slightly elevated land located approximately one-tenth of a mile west of the original rice fields. However, the settlement was abandoned by 1772 in favor of living quarters located in the upland pinelands. Geographer H. Roy Merrens and historian George D. Terry describe how the close proximity of inland rice plantation settlements to the rice fields represents early colonial perceptions of land use. "According to one [eighteenth-century] resident," cited the authors, "planters built their homes on the 'Edge of Swamps, in a damp moist Situation' because they wanted 'to view from their Rooms, their Negroes at Work in the Rice Fields.'" By gazing over developed agricultural spaces, planters viewed progress, order, and labor management that reflected the Enlightenment. They perceived this as a non-human world transformed from "savagery" to "civilization." However, early eighteenth-century colonists did not understand the connections between malaria-carrying *Anopheles* mosquitoes and low-lying habitats. In terms of this disease, the settlement pattern proved ill conceived and resulted in significantly higher mortality rates.[22]

While accurate statistics that document rural mortality rates are difficult to calculate, Merrens and Terry have sketched out a dismal scene for white and black residents. Approximately 37 percent of white males and 45 percent of white females born between 1721 and 1760 and surviving into adulthood in St. John's Parish died before their fiftieth birthday. Charleywood's Christ Church Parish offered more dire statistics, where 85 percent of all white males born between 1721 and 1760 and surviving into adulthood died before their fiftieth birthday. Malaria did not contribute to all of these cases, yet Merrens and Terry argue that 43 percent of

the interconnection between social and environmental formations on tidal rice plantations in *"What Nature Suffers to Groe,"* ch. 3.

[22] "The Following Lands," *South Carolina Gazette*; Matthias Rash to Peter Taylor, 18 March 1773, Taylor Family Papers, SCL; H. Roy Merrens and George D. Terry, "Dying in Paradise: Malaria, Mortality, and the Perceptual Environment in Colonial South Carolina," *Journal of Southern History* 50 (November 1984): 543–545, quote: 547. For understanding the intellectual framework of Enlightenment perceptions of the non-human landscape, see: William Cronon, "The Trouble with Wilderness; or, Getting Back to the Wrong Nature," in *Uncommon Ground: Rethinking the Human Place in Nature*, ed. William Cronon (New York: W.W. Norton & Co., 1995), 69–90, and Carolyn Merchant, "Reinventing Eden: Western Culture as a Recovery Narrative," in *Uncommon Ground*, 132–159.

recorded Christ Church deaths occurred within a four-month period between August and November suggesting an infectious cause.[23]

Planter families began moving away from the edges of rice fields and toward upland areas by 1760. One 1753 writer advised to build houses "at high places," where "things go better." High-ground settlements also avoided the seasonal freshets, which would inundate buildings located close to the floodplains. Thomas Hasell, of St. Thomas and St. Denis Parish, noted in 1722 that, "many families [were forced] to quit their houses and retire to the higher Lands in the woods" due to overflowing rivers and swamps. Albert Pouderous, rector in St. James-Santee Parish, commented in 1731 how a "water-flood that ruined" his parishioners' plantations "came six foot high into the houses." Wealthy planters began living on higher land and leaving plantations seasonally.[24]

By the 1770s, Beresford ordered two new settlements positioned in strategic locations. The upper Charleywood settlement, built in the Pamlico Terrace's sandy pine flatwoods community, was more than likely relocated for healthier living conditions. As Beresford was an absentee planter, the upper Charleywood settlement housed the plantation overseer and some select slaves. However, the other percentage of Charleywood's enslaved population had to endure exposed and sickly conditions at the second settlement located in the middle of the new rice fields. The centrally located Bay Hill settlement consisted of four houses, a corn house, a "mite pen," and a sick house. Bay Hill residents lived on an isolated stretch of high land approximately 100 feet wide and 460 feet long between the Fairlawn Canal and surrounding rice fields.[25] Bay Hill reflects the "separate residential zones" defined by anthropologist John Michael Vlach, which were conceived to divide the planter's family from their ever-increasing enslaved rice laborers. These zones, according to Vlach, "frequently [were] set miles away from the planter's residence; the quarters were sizable villages where slaves developed social routines of their own." Bay Hill was separated from the Charleywood settlement by the middle reserve and rice fields. This put slave dwellings closer to the

[23] Merrens and Terry, "Dying in Paradise," 541, 550.

[24] Walter L. Robbins, ed., "John Tobler's Description of South Carolina (1753)," *South Carolina Historical Magazine* 71 (July 1970): 158; Merrens and Terry, "Dying in Paradise," 549, quote: 547, n.37; Pouderous quotes in Arthur H. Hirsch, *The Huguenots of South Carolina* (Durham: Duke University Press, 1928; reprint, Columbia: University of South Carolina Press, 1999), 321.

[25] Matthias Rash to Peter Taylor, 18 March 1773, Taylor Family Papers, SCL; "A Plan of Charleywood Plantation," CCRMC.

work place and farther away from the upland residence. Moving slave settlements away from the big house, and closer to the work site, allowed enslaved laborers to travel from residence to work in less time.[26]

The efforts of inland planters to control water moved beyond their plantation borders and expanded into larger projects involving neighboring plantations or plantation communities. A movement of public and private canal initiatives took place throughout the eighteenth century to assure water for the effective transport of commodities and, simultaneously, to drain wetlands and flood their fields. With a majority of inland rice environments located on non-navigable waterways, landowners petitioned the proprietary and royal government for public canals for commerce. Inland planters also recognized the importance of these water highways to drain water from the broad low-lying wetlands. With the transformation of inland swamps into agricultural landscapes, those planter communities fanning out of Charleston began to construct canals on the Wando River and Biggin Swamp. Some requested the legislative assistance to fund internal improvement projects between 1719 and 1768 in Cooper River headwaters, Back River, the Ashley River headwaters, Caw Caw Swamp, and the northern branch of the Stono River.[27]

Daniel Ravenel's canal represented the extent and scope of private projects. Ravenel of Somerton Plantation constructed a mile and a quarter long canal in Biggin swamp. The canal was fifty-feet wide between his Somerton and Wantoot Plantations, powering his mill and providing manageable water flow to the inland rice fields. Also, the canal connected the four swamps into a single conduit and flowed into the Biggin Swamp Canal, a confluence of three canals stretching over thirteen miles through the surrounding wetlands. Ravenel's slaves constructed an artery between property holdings that was "traversed and intersected by dikes and tributary canals," and carved through "the dense growth" of Biggin's

[26] "A Plan of Charleywood Plantation," CCRMC; John Michael Vlach, *Back of the Big House: The Architecture of Plantation Slavery* (Chapel Hill: University of North Carolina Press, 1993), 155, 187; Chaplin, *Anxious Pursuit*, 262–271.

[27] George D. Terry, "'Champaign Country': A Social History of an Eighteenth Century Lowcountry Parish in South Carolina, St. Johns Berkeley County," (Ph.D. diss., University of South Carolina, 1981), 176–180; Petitions to the General Assembly, SCDAH; Thomas Cooper and David J. McCord, eds., *The Statutes at Large of South Carolina* (Columbia: A.S. Johnston), vol. 7 (1840): Act 410, pp. 479–480, mislabels Back River as "Black River"; Act 417, p. 481; Act 582, pp. 482–486; Act 672, pp. 492–496; Act 820, pp. 506–508; Act 969, pp. 513–519.

cypress swamp. Through the engineering of the canal, clearing of trees, movement of earth, and interaction with rice trunks and fields, Black Oak Agricultural Society's president Samuel Dubose considered Ravenel's canal "to have been the greatest work undertaken by a private citizen of this State up to the period of its construction."[28] Reflecting on the previous generation's innovation, Dubose wrote of Ravenel's plantations, "[it is] a source of surprise and wonder to examine the amount of labor and skill some in the [inland rice] fields in this neighborhood exhibit." Noting the broader complexity of the consolidation of water management by Biggin Basin inland rice planters, Dubose believed "to unite and concentrate these [canals] into one, and bear off the water when in excess, as well as distribute it into the fields of different plantations, called for judgment, perseverance and an amount of labor not easily understood."[29]

The private canal projects involved specific inland rice plantations and a small population that directly benefited from the improvements. Unlike the public improvements in navigation on creeks and rivers occurring throughout the eighteenth century, these new private canals did not benefit the greater population. Isolated to the few plantations bounding the watercourse, private canals improved navigation and irrigation for neighboring plantations of the planters involved. Also, the amount of slave labor needed was coordinated between the parties involved. Inland canals were wide enough for rice-flats transporting barrels of rice to Charleston, while in addition, the canals flooded or drained individual field divisions. Unlike tidal cultivators, who could flood and drain fields with the same canal utilizing the ebb and flow of the river, inland planters were limited to the downward flow of water. Planters could use these canals only to flood fields located below the canal or drain fields above the canal. For example, a mile-and-a-quarter-long drainage canal bisected Charleywood Plantation, creating a division between the upper and lower settlements and fields. The Fairlawn Canal stemmed from the neighboring Fairlawn Plantation and was wide enough for barges to transport rice and goods to navigable Guerin Creek, a tributary of the Wando River. In addition, the canal's central location provided irrigation to

28 Henry Edmund Ravenel, *Ravenel Records: A History and Genealogy of the Huguenot Family of Ravenel, of South Carolina*, (Atlanta: Franklin Printing and Publishing, 1898), 44–45.

29 Samuel DuBose, "Address Delivered at the 17th Anniversary of the Black Oak Agricultural Society, April 27, 1858" (Charleston: A.E. Miller, 1858), 9–10.

Charleywood's lower fields while draining water off Fairlawn's inland rice fields.[30]

Public canals benefited the larger planter community more than the private waterways. The public canal system added navigation and improved irrigation and drainage; these systems were larger in scope and their development involved organizing a slave labor force beyond that of a handful of planters. As historian Robert J. Kapsch notes, the colony's Assembly agreed to support these internal improvements, "based on sound economy ... because water transportation was faster, more reliable, and less expensive than road transportation."[31] The first movement of planters petitioning the Assembly to build canals benefiting inland rice planters occurred in 1740. Legislative acts in 1704, 1719, 1726, and 1734 had improved the navigation of creeks and rivers near inland rice fields, yet the 1740 Act specifically discussed how internal improvements would benefit residents by draining the land. The Act involved planters along the Jack Savanna, Horse Savanna, and Caw Caw Swamp, three drainage basins west of Charles Town that converged into Rantowles Creek, a northern tributary of the Stono River. In order to solve the problem of spring and fall freshets rendering the land "useless and unfit for planting and cultivation," planters petitioned the assembly to help open drains in these swamps. This was the first colonial effort to drain wetlands for agricultural purposes beyond individual properties. By identifying particular districts to drain, the 1740 Act addressed shifts in agricultural demands for improved land beyond isolated fields or specific plantations or farms.[32]

In response to the 1740 petition, South Carolina tapped into Rantowles planters' resources of enslaved labor and wealth. The Acts of 1740, 1754, and 1768 required enslaved laborers to dig a total of thirty-six miles of public drains in the Rantowles Creek Basin. This enormous amount of earth did not take into account the irrigation and drainage

[30] "A Plan of Charleywood Plantation," CCRMC; "Plan of Fairlawn Plantation in St. Thomas Parish, Charleston District," May 1794, no. 4339, John McCrady Plat Collection, CCRMC.

[31] Robert J. Kapsch, *Historic Canals and Waterways of South Carolina* (Columbia: University of South Carolina Press, 2010), 2.

[32] Act 672: "An Act to Cut and Sink Drains and Passages into the North and West Branches of Stono River," 1740, in Cooper and McCord, *Statutes at Large of South Carolina*, vol. 7: 492–496; Keith H. Beauchamp, "A History of Drainage and Drainage Methods," in *Farm Drainage in the United States: History, Status, and Prospects*, Miscellaneous Publication 1455, ed. George A. Pavelis (Washington, D.C.: United States Department of Agriculture, 1987), 15.

canals and embankments leading into the public canals. The Act's authors set guidelines for the canals "not to exceed fifteen feet in breadth." Eventually slaves moved approximately 17 million cubic feet of earth to construct these drains. To put this project into comparison, the Great Pyramid of Giza has a volume of 90 million cubic feet. To construct and maintain this intensive water diversion project, a hierarchy of planters served as commissioners for the related drains. Each of the commissioners had property bordering some part of the proposed canal route. The state assembly appointed three groups of commissioners to oversee the canals that flowed through the Rantowles Basin. The State "authorized and empowered" two groups of five commissioners each "to lay out, cut, sink, maintain and keep in repair" the Long and Horse Savanna drains. The Assembly charged a third group of three commissioners with the same responsibilities for the Caw Caw drain. In addition, each group was to assign "the following free drains or passages to carry off the said waters, at the proper costs and charges of the proprietors of the lands liable to be overflown thereby." By 1754, the assembly added an additional commissioner to the Caw Caw division because the drain expanded in length. Commissioners drafted the proposed watercourse and determined which plantations would benefit from the improvement project, "equally and indifferently to assess all the owners of the lands liable to be overflown by stopping up or diverting the waters or freshes of the said free drains, according to the number of acres subject to be overflown as aforsaid."[33]

Despite the State's contribution to improve drainage in the Rantowles Basin, local planters had to pay for the canal in two separate forms of tax. The first tax was monetary. Commissioners made "rates and assessments, in proportion to the number of acres each person is respectfully owner of." Planters paid the State a direct percentage for the number of acres drained from the canals. However, planters were taxed a second time as property owners had to devote all male slaves between sixteen and sixty years old to work on the project. Joyce Chaplin associates this type of labor to corvée, an obligation imposed on inhabitants of a district to

[33] Mileage was calculated by determining the limits of each passage described in Statutes and then compiling mileage from "A Plan of Several Tracts of Land Adjoining Each Others in South Carolina, Situate Part in St. Paul's, Part in St. Andrew's, Part in St. George's Parishes, 1775," Middleton Family Papers, Middleton Place, SC; "Map of the Roads in St. Paul's Parish, 1862," SCHS; Cooper and McCord, *Statutes at Large of South Carolina*, vol. 7: Act 672, pp. 492–496; Act 820, pp. 506–508; quotes: 492, 493; *State Gazette of South Carolina*, 31 October1785: 1; Chaplin, *Anxious Pursuit*, 264.

perform services, in this case with "slaves drafted to supply labor as a tax on the planters who stood to benefit from the completed project." Prior to this statute, the assembly has required property owners to provide their own labor to service projects; however, the reality was that property owners substituted slaves in their place. The Rantowles drains served as an early example of how the state taxed planters, through the use of enslaved labor, to complete the project.[34]

The Rantowles Creek drainage basin challenged aspiring rice planters. The basin is a low-lying Holocene back barrier and swamp deposit, set between upland Pleistocene beach deposits and the Suffolk Scarp. The drainage basin differed from the steep grades and definable floodplains of the previously discussed inland rice zones. Unlike the small-stream floodplains and second-order bottomlands, the Rantowles landscape consists of a series of pond cypress savannas that reflect the relatively flat and poorly drained land. High-pine communities formed upland fingers encroaching into the broad wetland basin. Freshets inundated this poorly drained basin, which made lowland cultivation difficult. The seasonal flooding washed away seedlings in the spring and damaged stalks in the fall. The downward flowing water also eroded any embankments in its path. Reverend William Guy, a rector in St. Andrew's Parish, reported in 1727 that travel remained difficult over the "two large branches of Stono River" because of "the bridges being broke down sometimes by the violent rains that happen there in the latter end of September." The basin had little grade or natural canalization; freshets swept across the broad basin until converging into Rantowles Creek.[35]

For planters, the Rantowles Basin topography provided a broad canvas on which to construct an intricate network of embanked fields and canals. The Pleistocene beach deposits created a barrier, serving as a natural embankment, between the basin and the Stono River. The only discharge point was where the Rantowles Creek blew out the geological barrier to connect with the coast. This barrier complex slowed the downward water flow, which translated into poor drainage during seasonal freshets. The 35,866 acres (56.04 sq. mi.) in the basin's watershed allowed great quantities of rainwater to flow downstream from the

[34] Act 672; Chaplin, *Anxious Pursuit*, 264.

[35] Porcher and Rayner, *Wildflowers of South Carolina*, 98–99; Rev. William Guy quoted in "An Account of a Parish, 1727/8," in *The Colonial South Carolina Scene: Contemporary Views, 1697–1774*. Tricentenial Edition 7, ed. H. Roy Merrens (Columbia: University of South Carolina Press, 1977), 84.

various ridges and "islands" dotted across the landscape, and funnel into the three Rantowles Creek tributaries. Despite the poor drainage, the Rantowles wetlands consisted of the type of soil that planters' desired for rice cultivation. The Rantowles Creek Basin consisted of a Mouzon-Brookman-Wahee soil association, defined as "somewhat poorly drained to very poorly drained soils that have a loamy surface layer and a loamy and clayey subsoil." A majority of the soil that supported the local rice culture was clay loam with "nearly level" declination toward the coast with slow water permeability or optimal water retention.[36]

Plantation names in this area represented the dramatic topography that people had to contend with in pursuing rice cultivation. For instance, "Clayfield" acknowledged the high clay content found within the plantation tracts in the Rantowles Creek Basin. A common nomenclature was the use of "hill." "Clay Hill," "Sandy Hill," " Poplar Hill," and "Walnut Hill," represented upland soil and vegetation associations, while "McPherson Island" and "Elliot's Island" described the highland knolls surrounded by lowlying wetlands. In an area inundated with water, highland knolls were just as valuable to settlements as the lowland swamps were to rice fields. As local historian Elise Pinckney stated, these names represented "the practice of celebrating any altitude" available to the planter. Samuel Galliard Stony, in describing the highland's importance a generation prior, noted, "any altitude is so precious that it is estimated almost in inches." These islands provided dry land needed for settlements dotting the basin. Bluffs and causeways linked these rural establishments. Bluffs provided planters with settings for upland crops while causeways served a dual purpose of containing or embanking water to rice fields, and providing roadways across low-lying wetlands.[37]

[36] McCartan, Lemon, and Weems, *Geologic Map of the Area between Charleston and Orangeburg*; Doar and Willoughby, *Pleistocene Dorchester and Summerville Scarps*; Jennie P. Eidson et al., *Development of a 10- and 12-Digit Hydrologic Unit Code Numbering System for South Carolina, 2005* (Washington, D.C.: US Department of Agriculture, 2005), 29; United States Department of Agriculture (USDA), *Soil Survey of Dorchester County, South Carolina* (Washington, D.C.: US Department of Agriculture, 1990), 62, 77. The USDA's Soil Conservation Service defines "nearly level" as less than 1 percent slope.

[37] Henry A.M. Smith, "The Ashley River: It's Seats and Settlements," *South Carolina Historical and Geological Magazine* 20 (April 1919): 85, 97–98, 108–115; Henry A.M. Smith, "The Upper Ashley: And the Mutations of Families," *South Carolina Historical and Geological Magazine* 20 (July 1919): 165–168; Pinckney and Stoney quoted in Elise Pinckney, "Still Mindful of the English Way: 250 Years of Middleton Place on the Ashley," *South Carolina Historical Magazine* 92 (July 1991): 151.

The Rantowles floodplain provided an uninterrupted expanse for enslaved laborers to carve geometrically ridged canals and embankments, unlike the narrow floodplains of early eighteenth-century rice fields. Tied into this created, intricate maze was the system of public drains. These fifteen-foot wide conduits created more irrigation and drainage options for regional planters, who routed water from them (or from intersecting tributaries) to irrigate their fields.

Both inland and tidal cultivation showed the importance of flowing water on and drawing water off the rice fields. Also, the grid-like field system used by both caused some confusion for later scholars, as the elaborate network of canals, ditches, drains, embankments, and dams became synonymous with both types of irrigation strategies. The confusion related to inland rice fields' evolving architecture, from their field boundaries defined by fluid upland topography to their geometrically rigid network of quarter ditches and drains. As inland rice networks expanded in acreage and infrastructure by the mid-eighteenth century, inland planters, like their tidal counterparts, intensified "compartmentalizing field sections" to control water on and off the fields. Planters' descriptions of "throwing water" and "letting off water" were the same. By the end of the colonial period, large inland fields had all the markers of their tidal counterparts, with the major exception that they did not use the rivers' ebb and flow. Large-scale rice cultivation, whether inland or tidal, relied on the massive infrastructure to control water and the labor force to carry out that requisite.[38]

Despite planters' organization and intent to construct and maintain a stable, extensive canal system, their drainage courses were not static. Major canals served as property lines and commissioners had to move plantation boundaries as the savanna streams changed route. Canals ebbed and flowed between neighboring planters' rice fields, as a consequence of alterations made to the path of the watercourse. As the public drains changed course under the direction of the commissioners, planters integrated the older canal system into their rice fields to compensate. Channels once used as central drains became secondary canals connecting the new public drains to the adjoining rice fields. Occasionally, planters would annex the neighboring fields if the fifteen-foot-wide canal moved onto the neighboring property. John Edwards, for instance, claimed

[38] Chaplin, *Anxious Pursuit*, 270–273; Norman R. Hawley, "The Old Plantations in and around the Santee Experimental Forest," *Agricultural History* 23 (April 1949): 86–91; quotes: Edelson, *Plantation Enterprise*, 105, 106.

twenty-six and one-half acres of John Miles' rice fields after commissioners moved the Horse Savanna public drain onto Miles' property. On this rerouted course, the canal changed directions three times. The original course had wrapped around a highland island. Commissioners decided to straighten the course by dredging a channel through the highland, yet afterwards the new alteration did not sufficiently move water. To augment passage, commissioners redesigned the drain, bypassing the knoll altogether and bisecting John Miles' rice fields. A similar scenario occurred downstream with Robert Miles and Edwards, where the commissioners straightened the public drain and this time severed Edwards' rice fields. The new channel placed one and six-tenths of an acre on Robert Miles' side of the canal, which the planter claimed as part of his property.[39]

With the colonial expansion of inland rice plantations, public and private drains for intricate field systems moved westward toward the Savannah River. For instance, planters on the headwaters of the Ashepoo River, southwest of Rantowles Creek and thirty miles from the coast, petitioned the Assembly to drain and make navigable the "Three Creeks" area. In 1742, planters petitioned only for drainage of Elliot's and Horse Shoe savannas, the eastern most tributary of the Ashepoo, but in 1756 they expanded their petition to include Chessy Creek, the middle tributary. Consistent with their work in the Rantowles Creek Basin, these planters sought to make the non-tidal tributaries navigable to transport their rice crop to market through watercourses.[40]

As irrigation management started to take hold along the tidal floodplains, planters with inland fields located on small-stream floodplains leading to tidal floodplains began to integrate inland and tidal irrigation. Plantations bordering tidal rivers had the ability to utilize a combination of irrigation practices, or "mixed regimes." Scholars have touched on mixed regimes while discussing variations in tidal rice irrigation.

[39] "A Plan of Several Tracts of Land," Middleton Family Papers; Plan, "At the Request of Dr. Haig Chairman of the Board of Commissioners Appointed by the Rest of the Legislature of This State for Opening a Canal for the Purpose of Draining the Swamps of Horse and Jack Savannas St. Paul's Parish Being Branches of the Stono River," May 1813, Middleton Family Papers, Middleton Place, SC; "A Plan of Land Belonging to John Edwards," March 1797, John McCrady Plat Collection, no. 1297, CCRMC; "Lands of Horse Savannah," May 1796, SCHS; "Plan of the Canal or Drain through the Middle of the Swamp of the Head of the Branch of the Stono River as Laid Out by Mr. Fenwick," SCHS.

[40] Cooper and McCord, *Statutes at Large of South Carolina*, vol. 7: Act 694, pp. 497–501; Act 844, pp. 510–512.

However, this scholarship has suggested that the process was simple, by describing planters damming swamp reservoirs on the edges of their tidal fields and using the impounded waters to supplement tidal river irrigation. In reality, planters used a variety of complex strategies to integrate the two irrigation practices. Some planters turned their older inland fields into tidal fields and used the original reservoir to supplement tidal irrigation during droughts. Other planters cultivated inland and tidal fields simultaneously, by transferring water through a canal system from the inland to the tidal fields when needed. In addition, the size of planters' labor forces and the relative productivity of existing inland field systems determined how much emphasis was placed on each irrigation method. Planters with financial and natural resources had the ability to work both field systems. Planters who had enough labor to divide between inland

FIGURE 3.3 Mixed regime rice fields, with inland rice fields located in the center, draining bottom right to Huger Creek and draining bottom left to Kensington Creek and Limerick tidal rice fields. Note the Windsor rice fields to the right.
"Plan of Limerick, a Plantation Belonging to Elias Ball, Esq." (detail), June 1797, John McCrady Plat Collection, 1541, Charleston County Register Mesne Conveyance, Charleston, SC

fields and tidal systems could maintain both agricultural zones. Planters who were strapped for labor tended to place a higher priority on their tidal fields.[41]

Conversion to these "mixed regimes" started to take place by the mid-eighteenth century. Henry Laurens instructed surveyor Alexander Gray in 1767 to "examine well [those plantations with] the flux and ebbing of tides [in close proximity to] Ponds, Lakes, or Creeks of fresh Water & its capacity of holding reserves of such Water, as well as of being easily & effectually drain'd in gluts of Rain, & be very minute in ascertaining how high in any of the Rivers & Creeks the water ever at any time run Salt or Saltish." To Laurens and other planters, reservoir-impounded water provided a valuable resource when tidal rivers were subject to salts. Droughts reduced the downward flow of fresh water that pushed toward the sea, allowing the "salt wedge" of brackish water to encroach upstream and damage the tidal irrigated crop. Relying on impounded inland water, planters reversed the irrigation direction, flooding rice fields on tidal floodplains from upland reservoirs.[42]

Topographic relations between existing inland reservoirs and tidal floodplains dictated how people connected the two systems. Henry Middleton's Ferry Tract Plantation consisted of inland rice fields later incorporated into tidal networks. The early fields paralleled the Combahee River bluffs, while the irrigation canals were connected to Bull's Creek, a tributary of the Combahee River, which separated the Ferry Tract from the northern Hobonny Plantation. The creek served as a drainage channel from the flooded inland fields to the tidal river, but it drained water efficiently only out of the embanked area during low tides. The spring-fed reservoir made up the central portion of these wetlands, flowing into inland fields on the upper and lower sections of the plantation. The Ferry Tract's upper fields relied on dual flanking canals drawing water on and off the fields. Trunk minders could fill or release water from ten individual fields without altering the adjoining quadrants. The dual system diverted water from an overflowing reservoir more effectively than a single canal. Unlike the basic inland model, the adjoining canals allowed water surges to bypass the fields and relieve any pressure on the reservoir

[41] For interpretations of mixed regimes, see: Hilliard, "Antebellum Tidewater Rice Culture," 98–100; Carney, *Black Rice*, 86–88; Edelson, *Plantation Enterprise*, 106–108; Groening, "Rice Landscape," 7778; Richard D. Porcher, *A Teacher's Field Guide to the Natural History of the Bluff Plantation Wildlife Sanctuary* (New Orleans: Kathleen O'Brien Foundation, 1985), 15.

[42] Groening, "Rice Landscape," 78; Hilliard, "Tidewater Rice Plantation," 62, 64.

dams. The success of this improved system depended upon the availability of larger labor forces to construct and maintain the canals. Plus, planters needed adequate suitable terrain to craft a canal system. Their enslaved laborers would have to construct an adjacent canal system with enough grade to efficiently drain water from the side of the fields. In comparison, the lower fields system relied on multidirectional canals, as the lower wetland topography was not subjected to the tight confines of the upland environment.[43]

Only a few feet in elevation determined what land Henry Middleton chose for inland fields and what land he devoted to cleared fields or wooded areas. Between the Combahee tidal lands and the inland rice fields lay Middleton's settlement and slave quarters. These buildings were situated on an island of high ground ten feet above the "low lands" and the rice fields on either side of the peninsula, connected only by a causeway to nearby roads. This finger of terra firma "cleared land," which varied between five and fifteen feet above sea level, defined his two inland field divisions. The cleared land was separated by "oakland uncleared" on the central portion of the plantation. These cleared highland areas were strategically divided into quadrants, much like the lowland rice tracks. Topography, in a sense, directed people to build living quarters and outbuildings on high land, and then surround the settlements with pastures, gardens, and fields.[44]

Elias Ball's Limerick and Comingtee Plantations represent the changing use of reservoirs to fit with changes in drainage systems and irrigation practices. A contemporary of Henry Middleton, Elias Ball cultivated land twenty miles northeast of Charleston. Limerick Plantation covered four biotic zones in the South Carolina Lowcountry. These zonesconsisted of longleaf pine forest, mixed hardwood forest, hardwood swamp, and freshwater marsh. The Ball family devoted each of the four terrains to rice cultivation. Slaves cleared the cypress, oak, hickory, and tupelo from the hardwood swamps before cultivating the land. Like Middleton's plantations, Limerick's reservoirs and fields maintained high water permeability from clay loam subsoil. Unlike the Argent series found at Middleton's Ferry Tract, Limerick's fields were situated on a highly fertile series called Meggett Loam. Noted by an early twentieth-century soil scientist during the waning years of the South Carolina rice

[43] "Map of a Plantation Belonging to Henry Middleton Esq., Called the Ferry Tract," May 1795, SCHS.
[44] Plat, "Ferry Tract," 26 January 1772, SCHS; http://nationalmap.gov/.

economy, Meggett Loam yielded forty to sixty bushels of rice per acre at the peak of the Lowcountry rice economy. The Balls connected these fields through a canal system to embanked land in the freshwater marsh to enable use of that land for rice cultivation. Like the Middleton property, settlements, gardens, and grazing land existed on high land.[45]

Limerick Plantation represents an inland rice field with intricate canal systems used simultaneously with tidal fields. These canals directed water from two reservoirs into specific inland rice fields, yet cultivators could move water toward the lower tidal fields if needed. The wider canals could direct larger volumes of water faster than smaller check banks. The two reservoirs varied in size in direct relation to the size of the impounded floodplains. Water in the first was retained by a dam across Kensington Creek, a tributary of the East Branch of the Cooper River. The impoundment formed a wide cypress swamp formed by Alligator Creek converging with Gough Creek. Kensington Creek and a reservoir formed the western boundary of Limerick. They served as a primary water source for the rice fields and a natural drain once the waters were released from the fields. This retention pond did not connect directly to the rice fields, so the Balls had to run an auxiliary canal to divert the water to Limerick. This additional canal system directed water from Alligator Creek into the plantation, following the low-lying land that became the rice fields.[46]

Limerick's success with the "mixed regime" irrigation came from trunk-minders' ability to direct water to specific inland and tidal fields. A twenty-three-acre reservoir, on the north side of the plantation, fed the inland field system. Gough Creek was part of a small-stream floodplain connected directly to the rice fields. Two adjoining inland rice fields lay beneath the reservoir's embankment; they funneled water through sixty-one acres until it was discharged into Kensington Creek or diverted to fifty-two acres of inland rice fields. This canal served the dual purpose of directing water from the fields and also irrigating portions of the lower

[45] William Lees, *Limerick: Old and In the Way, Archeological Investigations at Limerick Plantation, Berkeley County, South Carolina* (Columbia: Institute of Archaeology and Anthropology, 1980), 25–26; USDA, *Soil Survey of Berkeley County*, 6, 22; early 20th-century classification of Megget Loam was Portsmouth Clay Loam, see: W.J. Latimer, ed., "Soil Survey of Berkeley County, South Carolina" in *Field Operations of the Bureau of Soils, 1916* (Washington, D.C.: Bureau of Soils, 1921), 515.

[46] Cheryll Ann Cody, "Slave Demography and Family Formation: A Community Study of the Ball Family Plantations, 1720–1896," (Ph.D. diss., University of Minnesota, 1982), 92.

fields in this network. After the minders opened the trunks from these upper fields, the Limerick canal system directed the water to the lower fields if needed. The first direction of the upper canal system released the water back into Kensington Creek. The canal could also nourish the tidal fields along the East Branch of the Cooper River.[47] The second direction for the released water to follow was to another field system located right next to the East Branch of the Cooper River. This ninety-five-acre field system took on a completely different aesthetic compared to the upper network. The tidal fields had a consistent geometric shape and did not follow the natural boundaries of knolls or bluffs. These fields existed on the flood plains of the Cooper River tributary in a fashion similar to Beresford's Charleywood rice fields along Guerin Creek. The broader floodplains enabled these planters to design larger divisions with fewer topographical restrictions. Embankments separated the fields from the ebb and flow of the river as canals leading from the upper fields provided the water source for these lower fields.

Comingtee Plantation provides an example where planters abandoned inland fields, yet still used old reservoirs for supplemental irrigation for their emerging tidal systems. Elias Ball's Comingtee served as the family seat, located at the confluence of the Eastern and Western Branches of the Cooper River. Elias' father, Elias "Red Cap," inherited the tract from his great-aunt, Affra Harleston Coming, in 1698. Affra and her husband John received land from headrights by importing indentured servants between 1671 and 1678. Originally named "Coming's T" after the T-formation of the eastern and western branches converging to form the Cooper River, the plantation name evolved into "Comingstee" and eventually "Comingtee." Elias "Red Cap" began his plantation enterprise producing a variety of standard products at the turn of the eighteenth century, most commonly tar, pitch, corn, and rice. In 1722, he produced 102 barrels of rice with forty-three slaves at his three plantations: Comingtee, Dockum and Cypress Grove. By 1740, Elias "Red Cap" produced a substantial 223 barrels of rice with up to 100 slaves at Comingtee. The inland fields were irrigated by reservoirs that formed Comingtee Creek. Despite the substantial crop in 1740, Comingtee's output was inconsistent due to freshets and drought. Between 1732 and

[47] "A Plan Exhibiting the Shape and Form of a Body of Land Called Limerick," March 1786, Ball Family Papers, USC; "Plan of Limerick, a Plantation Belonging to Elias Ball, Esq.," June 1797, John McCrady Plat Collection, no. 1541, CCRMC.

1742, Ball produced an average of 100 barrels annually from Comingtee's inland fields.[48]

The son of Elias "Red Cap" abandoned Comingtee's inland fields by 1775 to focus on the tidal floodplains. By 1790, Ball was planting 225 tidal acres at Comingtee. Noting the delicate balance between too much and not enough rain, Ball mentioned the "grate part of it seems to luxuriant and my reserve fine and full of water." The adequate water would have served him well later in the summer if a strong flow of brackish water had inundated the Cooper. The skill of managing impounded inland reservoirs to balance tidal surges on the Cooper floodplains became fully apparent during freshets. Freshets breached the Comingtee "great dam," which had been in place since 1764, flooding the downstream rice fields and causing "total ruin" that "carried away everything before it." For planters like Ball, inland reservoirs provided an advantage during droughts, yet they could inundate the fields if, in addition, there were floodwaters. Unlike Ball's Limerick system, this operation at Comingtee did not utilize a washway canal to relieve pressure on the reservoir dam. At Comingtee, the plantation sat between the Bethera Scarp and the Cooper River, where five creeks converged into Comingtee creek. The downward flow of water, on a moderately graded elevation, with a twenty-foot decline in less than a quarter of a mile, put more pressure on the reservoir impoundments compared to other locales with less grade.[49]

To counter the freshets' downward force, Elias Ball constructed a string of five reservoirs dotting Comingtee Creek like a necklace of pearls. The five reservoirs impounded water in strategic topographical locations. At those sites, enslaved laborers constructed retaining dams that were

[48] Elias Ball Account and Blanket Book, 1720–1778, Ball Family Papers, SCHS; Elias received the nickname "Red Cap" by his decedents to distinguish the elder Ball from later namesakes. "Red Cap" described the cap that Elias wore in a portrait. Edward Ball, *Slaves in the Family* (New York: Farrar Straus & Giroux, 1998), 34–35, 38–42, 96, 176; Cheryll Cody calculated that Elias Ball was specifically purchasing enslaved Gambians by the 1750s and 1760s. Seven of the eight men between the ages of 30 and 34 and two men between the ages of 45 and 47 were Gambian. Cody, "Slave Demography and Family Formation," 44, 49.

[49] "Elias Ball, Limerick, to Elias Ball, Bristol, England," 6 June 1790, Ball Family Papers, SCHS; "A Plan of Comingtee Plantation," May 1786, CCRMC: http://nationalmap.gov/; Ball, *Slaves in the Family*, 35–42. Keating S. Ball also records the Comingtee Big Dam breaking and washing away the rice crop. He records the repaired dam having a width of fifty feet at the bottom, eleven and one half feet at the top, and a length of 100 feet. Keating S. Ball Plantation Book, 8 August – 17 December, 1849, John and Keating S. Ball Books, vol. 5, UNC.

connected by simple canals to the next reservoir downstream. Big Dam Reserve, the largest reservoir, was also the oldest and most problematic of the impoundments. Nicknamed "Ball's Folly," this reservoir was the first impediment against the Comingtee Creek freshets and subject to the most common breaches. "Daniel's Dam" was the second reservoir and consisted of "calm water and moss hung cypress." The "Bridges Reserve" was the third reservoir in the Comingtee system, formed by a causeway leading to the Ball's So-Boy Plantation. A short canal connected Bridges Reserve to "Rainy Basin," which Ball family historian Anne Simons Deas described as a "lively little reserve." Another canal drained water to the fifth and final reserve, called "Cork Gate," named after the large flood gate feeding water into the rice fields.[50]

Ball's transformation of inland rice fields into reservoirs reflected how planters creatively managed the plantation infrastructure to best meet their needs. Ball assigned much labor and energy "securing the leads" of his former inland fields to hold back water pressure enhanced by the decline in grade. However, freshets inundating these reserve and canal systems ultimately had no place to go. Nonetheless, the steep grade was advantageous providing definable highland close to the Cooper River floodplains. The Comingtee settlement sat on a twenty-foot knoll less than one-half mile away from the navigable river; this directly contributed to the plantation's early settlement by 1680. Also, the narrow watershed provided planters, like John Coming or Elias "Red Cap" Ball, with evident irrigation zones when inland rice cultivation took place on the property. The narrow inland floodplains, when combined with steep grades, enabled early rice cultivators to dam reservoirs efficiently and construct basic rice fields. However, with the advent of tidal technology and the Ball's creative use of multiple irrigation sources to maximize cultivation output, the system worked too well. The expanded reservoirs inundated the original rice fields with more water than they could retain, causing havoc during freshets.[51]

50 Anne Simons Deas, *Recollections of the Ball Family of South Carolina and the Comingtee Plantation* (Summerville, SC: Alwyn Ball, Jr., 1909), 16–18, 117–118; USDA, *Soil Survey of Berkeley County*, 47.

51 "Elias Ball, Limerick, to Elias Ball, Bristol, England," Ball Family Papers, SCHS; "A Plan of Comingtee Plantation," May 1786, CCRMC: http://nationalmap.gov/; Ball, *Slaves in the Family*, 35–42; Stanly South and Michael Harley, *Deep Water and High Ground: Seventeenth Century Low Country Settlement*, Research Manuscript Series 166 (Columbia: University of South Carolina Institute of Archaeology and Anthropology, 1980), 4–6, 24–35; Deas, *Recollections of the Ball Family*, 27. A similar situation occurred on

Planters began associating inland rice cultivation by the 1750s with myriad natural disasters, reflected in disease, declining soil fertility, pests, and unpredictable water supply. This led to its decline in popularity by the end of the American Revolution. By the turn of the century, colonists and slaves had spent most of their time on or near the inland wetlands and they began to associate death and mortality with this vast rice environment. In contrast, optimistic wealthy colonists viewed the lush wetland environments as fertile rice grounds with bountiful opportunity. Dr. George Milligen wrote in the early 1760s that the inland and tidal rice environments were the "Golden Mines of Carolina." The rich "cypress, river, and cane swamps [were the] source of infinite Wealth, and will always reward the industrious and persevering planter." In reality, enslaved African Americans served as the "miners," and in the same way that hazards presented themselves in mine shafts, so disease and death found themselves in the rice fields. So here lies the contradiction of owners' views toward their slaves. Planters valued their laborers' financial value and expense up to a point, yet knew – ultimately – the enslaved laborers were there to grow rice, to make the planter money, at any expsense.[52]

Despite the financial rewards obtained from rice cultivation, planters' perceptions of inland environments had dramatically changed, once again, by the eve of the Revolutionary War. Impounded reservoirs and rice fields created optimum environments for malaria carrying *Anopheles* mosquitoes. As colonists developed more inland rice fields and settled more land in the Carolina frontier, the malaria vector migrated with them. Beginning in spring as mosquitoes hatched in water-drenched fields and multiplied through the summer, malaria spread from host to host, transferred by the mosquito, bringing epidemic fevers, aches and chills to the rural residents, with potential for chronic relapses.[53]

By the 1730s, the Charles Town area became "a scene of diseases." The warm wet climate created a suitable environment for insect vectors

the neighboring Rice Hope Plantation, where a 1851 plat listed thirty acres as "Inland Rice Swamp, Now a Reservoir," "Plat of a Tract of Land Called Rice Hope," January 1851, John McCrady Plat Collection, no. 1512, CCRMC.

[52] Merrens and Terry, "Dying in Paradise," 540; George Milligen-Johnson, *A Short Description of the Province of South Carolina, with an Account of the Air, Weather, and Diseases, at Charles Town, Written in the Year 1763* (London, 1770), 4; Morgan, *Slave Counterpoint*, 33.

[53] Peter McCandless, *Slavery, Disease, and Suffering in the Southern Lowcountry* (New York: Cambridge University Press, 2011), 45–46.

while immigrants imported potential hosts from Africa, Europe, and the Caribbean. Smallpox and yellow fever became urban epidemics, as both spread quickly and furiously through the condensed human population. Malaria, according to medical historian Peter McCandless, was endemic in the Lowcountry, the "equivalent of background noise." McCandless states that malaria was "regularly present but rarely caused a ripple in the public sphere" as its effects became part of Lowcountry residents' daily lives. Despite the lack of public fear, malaria took its toll on rural residents constantly exposed to large populations of mosquitoes. The primary local vector, *A. quadrimaculatus*, thrived in the inland rice environments, as the *Anopheles* mosquitoes preferred sun and standing water for their breeding sites. Inland reservoirs, coincidently, provided an ideal environment for these vectors.[54]

When the connection between sickness and inland rice fields became evident, colonists began to "reappraise their environment," reverting back to seventeenth-century perceptions of wetlands as "dismal."[55] By the mid-eighteenth century, colonists recognized "low, flat parts of the country, and such as lie in the vicinity of the rice and indigo swamps, as well as the bad quality of water in such situations, render some parts of those countries unhealthy." Colonists, unaware of disease biology and ecology, relied on sensory observations to understand their environment. Odorous miasmas emitted from swamps were one way for residents to explain the unhealthiness associated with the low-lying rural environment. As described by the anonymous author of *American Husbandry* in the late colonial period, "from the mud of these stinking sinks and sewers the heat exhales such a putrid effluvia as much necessarily poison the air."[56] Disease also appeared from "sudden changes in climate" that resulted in "excessive heat." Unaware of the connection between mosquitoes and malarial infections, Lowcountry residents correctly deduced that inland rice fields were unhealthy environments in the

[54] McCandless, *Slavery, Disease, and Suffering*, 39–43, quotes: 39, 40; Timothy Silver, *New Face on the Countryside: Indians, Colonists, and Slaves in South Atlantic Forests, 1500–1800* (Cambridge, NY: Cambridge University Press, 1990), 160, 161–163.

[55] Merrens and Terry, "Dying in Paradise," 549; Ann Vileisis, *Discovering the Unknown Landscape: A History of America's Wetlands* (Washington D.C.: Island Press, 1997), 33.

[56] William Barton, "Observations on the Probabilities of the Duration of Human Life, and the Progress of Population, in the United States of America: In a Letter from William Barton, Esq. to David Rittenhouse, L.L.D. President, A.P.S." *Transactions of the American Philosophical Society* 3 (1793): 49; *American Husbandry* quoted in McCandless, *Slavery, Disease, and Suffering*, 18–19.

warmer months. Smell and temperature became gauges for quality of living conditions, and planters began in the late colonial period to relocate their summer homes away from the inland rice plantations and to urban centers, coastal islands, or upland settlements.[57]

The real and perceived association of inland rice fields and unhealthy environments contributed to this shift in settlement patterns. As planters accumulated more capital at the hands of expensive enslaved laborers, they had increasing opportunities to purchase more land further into the Carolina frontier. Planters began establishing family seats at plantations located on dry land deemed healthy, while outlying properties became agroecological landscapes. They escaped the seasonal malaria outbreaks by traveling to pineland villages, coastal hamlets, or European cities. Once cooler weather settled into the Lowcountry, a reverse migration took place with planters departing from their retreats and returning to the family seats. "Time and experience had now taught the planters," wrote an eighteenth-century observer, "during the autumnal months, their living among the low rice plantations subjected them to many disorders, from which the inhabitants of [Charles Town] were entirely exempt."[58]

By the mid-eighteenth century, inland planters made distinctions between environments that were best suited for agriculture and those best suited for homesteads. For example, plantations along the Ashley River were in close proximity to Charles Town where planters built their houses on bluffs to escape the devastating effects of malaria. However, this land did not support the best rice production. Although inland rice gained a solid foothold in this area during the eighteenth century and enabled successful planters to acquire better-suited lands, the Ashley River could not provide an effective transition to sustain the more profitable tidal culture.

The relocation of enslaved laborers away from the older inland plantations reflects the changing settlement patterns occurring through the

[57] Barton, "Observations on the Probabilities of the Duration of Human Life," 48; H. Roy Merrens, ed., "A View of Coastal South Carolina in 1778: The Journal of Ebenezer Hazard," *South Carolina Historical Magazine* 73 (October 1972): 187; Walter L. Robbins, ed., "John Tobler's Description of South Carolina (1753)," *South Carolina Historical Magazine* 71 (1970): 158; Merrens and Terry, "Dying in Paradise," 549. For a sensory history of wetlands, see Conevery Bolton Valencius, *The Health of the Country: How American Settlers Understood Themselves and Their Land* (New York: Basic Books, 2002), 114–119, 145–152, and Vileisis, *Discovering the Unknown Landscape*, 33–37.

[58] Merrens and Terry, "Dying in Paradise," 549; William Barton, "Observations on the Probabilities of the Duration of Human Life," 49.

Lowcountry. For example, the slave population in St. James-Goose Creek declined 10 percent between 1745 and 1790, although the number of free families increased 50 percent. During this forty-five-year period, the number of families owning fifty or more enslaved people in St. James-Goose Creek declined from 28 percent to 10 percent. The shift in numbers reflects how the parish transformed from a cradle of inland rice cultivation to enclaves of ornamental family seats.[59]

The Ashley River does not have a large watershed. It originates in the Great Cypress Swamp and stretches only thirty miles, relatively short compared to sprawling Edisto and Santee rivers that originate at the base of the Blue Ridge Mountains. It is more of an arm of the sea instead of a tributary to the ocean: brackish water flows further up the Ashley than any other tidal rivers. This made successful tidal rice cultivation very difficult. Compared to other rivers, first- and second-order streams connecting to the Ashley River were not in abundance and this made it difficult for planters to make substantial yields in inland cultivation.[60]

In addition to unhealthy landscapes, declining soil fertility diminished the role of inland rice culture in the South Carolina market economy. With the constant efforts of inland planters to manipulate soil and water for higher agricultural yields, a series of problems in cultivation became apparent. Both the market's increasing demand for Carolina rice by the 1760s and correlating intensive agricultural practices led to inland soil exhaustion. Rice extracts nitrogen from the soil and planters would see diminishing returns if the land was not fertilized. Unlike tidal rice fields that enjoyed natural fertilization from river sediment, inland irrigated fields received limited amounts of natural soil deposits. This difference occurred because reservoirs trapped downward flowing alluvium and suspended organic material, depleting nutrients in the water released into the fields. As early as five years after consecutive planting, rice planters noticed decreases in their fields and sought other options, such as clearing new inland wetlands or transferring labor to the emerging tidal floodplains.[61]

[59] Philip D. Morgan, "A Profile of a Mid-Eighteenth Century South Carolina Parish: The Tax Return of St. James', Goose Creek," *South Carolina Historical Magazine* 81 (January 1980): 53–56.

[60] Bill Marshall, *Ashley Scenic River Management Plan* (Columbia: South Carolina Department of Natural Resources, 2003), 2.

[61] Silver, *New Face on the Countryside*, 168; Edelson, *Plantation Enterprise*, 139; Converse D. Clowse, *Economic Beginnings in Colonial South Carolina, 1670–1730* (Columbia: University of South Carolina Press, 1971), 127.

By the final decades of the eighteenth century, inland rice cultivation had a reputation for "poor soil." Inland rice swamps did not have the "inexhaustible fertility" that colonists described decades before. Alexander Hewatt declared that the central concern of inland rice planters was "the art of making the largest profit for the present time." "[A]nd if this end is obtained," he noted, "it gives them little concern how much the land may be exhausted." Hewatt's criticism stemmed from the poor husbandry practices taking place by the end of the colonial period. Despite Hewatt's critique, he represented the minority of planter opinion. With expansive landholdings, successful rice planters could move to new fields or plantations, abandoning old fields in the process. "The emulation that takes place amoung the present generation, is not who shall put his estate in the most beautiful order, who shall manage it with the most skill and judgment for posterity," wrote Hewatt, "but who shall bring the largest crop to market. Let their children provide for themselves." Thomas Smith, property manager of Charleywood, expressed this frustration when writing to the new absentee owner of this plantation in 1773. Smith informed Peter Taylor, who purchased the plantation from Richard Beresford's estate that year, that Beresford "was a very Slovenly planter and hurted the land he planted so we must clear as much new land as we can."[62]

Although inland plantations were not entirely monocrop regimes, by the late-colonial period rice cultivators experienced problems with pests commonly associated with intensive agricultural practices. The narrow tree-lined inland corridors shaded rice fields and decreased crop output. Josiah Smith informed absentee planter Samuel Waddingham "that your Rice was but indifferent ... being so much shaded by the cypress and other trees in the swamp." Caterpillar epidemics occurred on inland fields, which affected neighboring plantations on the Ashepoo River headwaters. Planters were forced to release precious impounded water to flood the fields in order to kill off the pests feeding on the crop, while enslaved field hands beat the caterpillars off the top of the plant above the waterline. In 1768, Peter Manigault described the devastation at Caw Caw Swamp where freshets were destroying what little crop remained from caterpillar infestation. One year later, Manigault's Goose Creek plantation suffered substantially. He lost 80 percent of its eighty-acre

[62] Douglas C. Wilms, "The Development of Rice Culture in Eighteenth Century Georgia," *Southeastern Geographer* 12 (January 1972): 49; Alexander Hewatt, *An Historical Account of the Rise and Progress of the Colonies of South Carolina and Georgia*, II, 305–306; Thomas Smith to Peter Taylor, 14 July 1773, Taylor Family Papers, USC.

crop. The fields were "covered with thousands of Grasshoppers which sucked the Grain & prevented it's filling"; they were replaced by a "New Species of Bugs that eat the Blossoms of the Rice & prevented its filling" and then they endured "a small Insect like an Ear Wig equally destructive with the last."[63]

Rice as an agro-industry also attracted migratory birds feeding off the grain, such as the bobolink, red-winged blackbird, and crow. Planters especially feared the bobolink, called the "ricebird" or "maybird," whose large migratory groups would devour rice crops throughout the Lowcountry in the late spring and early fall. Rice planter J. Motte Alston observed they "come in myriads and in flocks so dense as to cast a shadow on the green and golden fields." As one correspondent wrote in 1750, "with regards to the *Rice-Birds*, it is almost incredible what devastation these little creatures will make." This bird had such a reputation for feeding on the grain, that the bird's name, *Dolichonyx oryzivorus*, means "rice eating." Modern research reveals 76 percent of their food is rice.[64]

Passing through the Lowcountry on their northward migration in late-April to late-May, and returning again in mid-August to late-September, the birds' migratory pattern coincided with stages of rice sprouting. The birds fed on the tender rice seedlings during their spring migration north, only to feed on the "soft and milky" grain during their fall migration south. Planters were forced to alter their planting schedule to avoid having rice in the fields when bobolinks were passing through the area because of the potential loss of crop to these birds. Mark Catesby recounted one unprepared Ashley River planter losing forty acres of his fall harvest to rice birds, only to question, "whether what [the rice birds] left was worth the expense of gathering in." Usually planters would start cultivating rice earlier to synchronize flood stages with ricebird

63 Josiah Smith to Samuel Waddingham, 22 July 1773, Josiah Smith to George Austin, 30 July 1772, Josiah Smith, Jr., Lettercopy Book, 1771–1784, Southern Historical Collection, University of North Carolina; Maurice A. Crouse, ed. "The Letterbook of Peter Manigault, 1763–1773," *South Carolina Historical Magazine* 70 (April 1969): 96 and (July 1969): 179.

64 Brooke Meanley, *Blackbirds and the Southern Rice Crop, Resource Publication 100* (Washington, D.C.: United States Department of the Interior, Fish and Wildlife Service, 1971), 39; Alexander, Sprunt, Jr., and E. Burnham Chamberlain, *South Carolina Bird Life*, rev. ed. (Columbia, SC: University of South Carolina Press, 1970), 489; Porcher, *Teacher's Field Guide*, 20; Alston quote found in Childs, *Rice Planter*, 76; "Devastations by Rice Birds. Charles Town; S. Carolina, Oct. 15," *Gentleman's Magazine* 1 January, 1751: 10.

migrations.[65] Colonial accounts, however, speculate that ricebirds damaged between 30 and 40 percent of the rice crop. Cultivators were forced to divert substantial work hours to ward off these birds during the growing season. Enslaved people, or "bird-minders," were placed in the fields with muskets and whips to scare these pests away from the rice crop. The correspondent for the *Gentleman's Magazine* stated in 1750 that, "2, 3, or more negroes are constantly kept traveling from the time the rice begins to ear, until it is full enough to cut, through every rice field, up to their knees and waits in water, continually hallowing and beating any sounding things to keep these birds from alighting there on."[66]

Ironically, planters relied on ricebird migration to help combat other threats to the crop: rice worms, grubs, and maggots that fed on the rice stalks and roots. The two-week flooding period that deterred ricebirds from settling upon the fields, helped also to kill off worms and grubs; yet a poorly drained field promoted maggots. Fine-tuning the water control of rice fields relied on a constant balance between migrations, habitats, and human observation. Overseers had to time flooding properly to protect the grain from ricebirds, yet use these avian pests to eradicate any insects that spawned in the flooded fields.[67]

In addition to malaria, reduced soil fertility, and pests, unpredictable water supply provided the final issue that challenged landscapes on inland rice plantations. Even with an intensive focus on controlling water through rice fields during the eighteenth century, inland planters still suffered from freshets and droughts. The unpredictability of downward flowing water, dependent on weather patterns, created precarious situations for growing rice and fostered the loss of productivity. Freshets breached reservoir dams and embankments, forcing enslaved field hands

[65] Catesby quoted in "Of the Rice–Bird," *Gentleman's Magazine* January 1, 1751: 11. Ricebirds are documented in Stewart, "*What Nature Suffers to Groe*," 161–162; Richard D. Porcher, "Rice Culture in South Carolina: A Brief History, the Role of the Huguenots, and the Preservation of Its Legacy," *Transactions of the Huguenot Society of South Carolina* 92 (1987): 7; Silver, *New Face on the Countryside*, 152; Francis Harper, ed., *The Travels of William Bartram: Naturalist's Edition* (New Haven: Yale University Press, 1958; reprint, Athens: University of Georgia Press, 1998), 188. Heyward notes that "Never was any planting done between the middle of April and the last of May. The reason for not planting after the tenth of April was the coming of the May birds, which, during the month of May, were on their way northward, after wintering in the far South, and could always be depended upon to appear in our rice fields. They seemed to travel on a regular schedule, and were always on time." Heyward, *Seed from Madagascar*, 31; "Devastations by Rice Birds," 10.

[66] Heyward, *Seed from Madagascar*, 32; Childs, *Rice Planter*, 76; Doar, *Rice and Rice Planting*, 27; "Devastations by Rice Birds," *Gentleman's Magazine* 1 January, 1751: 10.

[67] Doar, *Rice and Rice Planting*, 27–28; Porcher, *Teacher's Field Guide*, 19.

to repair the earthen structures and plantation carpenters to build new trunks. Henry Laurens witnessed "the most violent of storms" that had "drowned all the low inland swamps and ruined the growing Rice." He noted how Ralph Izard's three Jack Savannah plantations "have suffered as much as any." The "incessant rains" carried off seedlings before the plants took root, and would "damage" the rice stalk once it reached a mature stage. Droughts, on the other hand, prevented aggressive field flooding. The lack of flooding led to more invasive weeds in the rice fields; this meant field hands would spend more of their time removing weeds with less time available for nurturing crops.[68]

Observing eighteenth-century inland planter journals, one can get a grasp of the importance of weather to these people. Weather observations and their effect on the planters' crops remain consistent through these primary sources. The first weeks of planting rice were "a crucial time and require[d] the vigilance and judgment of the planter, for heavy rains, or severe droughts prove equally fatal, and put him in the necessity of re-planting," explained Timothy Ford. "In one case the banks around his field must be opened, and every possible drain made use of to draw off the incumbent water," Ford observed, "in the other case the sluices must be opened from the reservoirs and the water brought upon the field taking care that it remain not too long-ordinarily from 6 to 18 hours." Summer droughts prevented reservoirs from recharging "when there was no water in any of the creeks." Planters revealed in their private journals how the lack of reserve water forced them to abandon their flood schedule in the fields, where "everything was perishing" from the intense Lowcountry heat. Alice Izard, writing to her daughter in New York at the turn of the nineteenth century, commented on a passing storm "raining very seriously, to the great joy & comfort of the inland Rice Planters." Drought, too, was welcomed by some planters. Peter Manigault expressed that the "great drought" between March and May 1764 "was of Service to the Rice" giving relief from the spring freshets.[69]

---

[68] Josiah Smith to Samuel Waddingham, 22 July 1773, Josiah Smith, Jr., Lettercopy Book, 1771–1784, UNC; Ellen Shlasko, "Carolina Gold: Economic and Social Change on a South Carolina Rice Plantation," (Ph.D. diss., Yale University, 1997), 87; Henry Laurens, "An Original Letter of Henry Laurens," *Historical Magazine* 10 (April 1866): 99.

[69] Ford, "Diary of Timothy Ford," 182; Diary and Account Book of Rene and Henry Ravenel of Hanover, 15 and 29 September 1752, 22 August 1753, 25 August 1755, 10 May 1760, 19 October 1771, Ravenel Family Papers, SCHS; Alice Izard, Charleston, to M.I. Manigault, New York, 11 July 1801, Manigault Family Papers, 1750–1900, USC; Crouse, "The Letterbook of Peter Manigault," 84, 90. Also see: Stewart, "*What Nature Suffers to Groe*," 155–157.

The story of inland rice cultivation during the colonial period is one of changing environments, both real and perceived, and how people reacted to those changes. As planters expanded the built environment by intensifying their water management systems – through reservoirs, canals, and embankments – they inadvertently made living and working conditions inhospitable. Planters and slaves creatively constructed inland plantations in a variety of environments, but in doing so they put themselves in the way of freshets, disease, and pests. "We can see what the planters at first did not," observes Peter McCandless, "namely that malaria was an enemy they had themselves largely created." McCandless' statement is relevant to the other problems associated with inland rice culture. Agricultural problems commonly attributed to nature were actually human induced. Also in an effort to maximize crop output through available technology and labor, planters created an altered environment that resonated in changes in settlement patterns and deteriorating working conditions. By the eve of the American Revolution, this altered environment was almost void of free whites, while their enslaved laborers suffered to cultivate the cash crop fueling the Lowcountry plantation economy. The inland plantation landscape, however, would soon change dramatically.[70]

In addition to the series of natural disasters taking place on late-colonial inland rice fields came war: the American Revolution provided the definitive period of decline for the reservoir culture. With eight years of warfare leaving untended rice fields destroyed or overgrown, much of the plantation infrastructure was left to ruin. Freshets eroded reservoir dams and field embankments, while "volunteer," or wild, rice overtook uncultivated plots. Enslaved African Americans saw an opportunity to depart plantations, leaving only a limited labor force to continue cultivating rice. After the Revolution, agricultural change rippled through the Early Republic period as Lowcountry planters sought more capital to reinvest in their neglected plantations. As a result of new opportunities, many planters replaced inland for tidal rice and indigo for cotton.[71]

70 McCandless, *Slavery, Disease, and Suffering*, 45. On the complexity separating the human and natural environments, see Stewart, "*What Nature Suffers to Groe*"; Mark Fiege, *Irrigated Eden: The Making of an Agricultural Landscape in the American West* (Seattle: University of Washington Press, 1999); and Ted Steinberg, *Acts of God: The Unnatural History of Natural Disaster in America* (New York: Oxford University Press, 2000).

71 "Affidavit by Benjamin Smith on account of Peter Taylor, Whitehall," 7 July 1785, Taylor Family Papers, USC; Chaplin, *An Anxious Pursuit*, 236–237; Morgan, *Slave Counterpoint*, 60, 61, 64.

4

"To Depend Altogether on Reservoirs"

Upper Wando River Rice Cultivation, 1783–1860

Nestled in the battleground between encroaching suburban development and conserved Francis Marion National Forest exists a string of former rice plantations dotting the headwaters of the Wando River. As one hikes along the established trail-systems from the sandy upland pine stands down to cypress communities or *spartina* fields (both signifying the heavy dependence upon standing water) one notices the grid-like formations associated with Lowcountry rice plantations. Upon reaching the destination, one is greeted with miles of embankments that stretch into the horizon – a testimony to the toil and labor produced by the enslaved that serves as a permanent physical symbol of their legacy. A monument to the tremendous energy injected into growing a single crop. To the acute observer, these rice fields have the scope of many of the preserved tidal rice fields dotting the Lowcountry. The height of the embankments, dimensions of the impounded fields, width of the drains and canals all have similar characteristics. Even the aesthetics of these upper-Wando fields, in relation to the broad and meandering tidal river, closely resemble the tidal field-systems and their connection to freshwater tidal hydrology. Despite these similarities, the close understanding of this microenvironment reveals a different story that highlights how planters and slaves had to work within the limitations of this topography to grow the desirable cash-crop.

Although tidal rice plantations dominated the landscape through the antebellum period, select inland plantations continued to thrive and produce successful agricultural returns. In 1850, the two most successful rice operations in Christ Church Parish, east of the Cooper River, were inland rice plantations. Despite the continued practice of inland rice cultivation after the Revolutionary War, current historiography has overlooked the role inland rice culture played in the first half of the nineteenth century. Scholars have produced extensive analyses of the political, cultural, economic, and environmental themes surrounding tidal rice agriculture, with illustration of inland practices only providing an introduction to their larger arguments.[1] One reason for this disparity is the availability of source materials. A wide variety of primary sources exist that document nineteenth-century tidal rice plantations, and these contribute to historians' focus on this plantation system. Multiple volumes of papers and plantation journals described the process of tidal rice cultivation, leaving bountiful information of planter management and labor formation.[2] With a plethora of sources directing scholars

[1] David Doar, *Rice and Rice Planting in the South Carolina Lowcountry* (Charleston: Charleston Museum, 1936; reprint, 1970); Duncan Clinch Heyward, *Seed from Madagascar* (Chapel Hill: University of North Carolina Press, 1937; reprint, Columbia: University of South Carolina Press, 1993); Joyce E. Chaplin, *An Anxious Pursuit: Agricultural Innovation and Modernity in the Lower South, 1730–1815* (Chapel Hill: University of North Carolina Press, 1993); Daniel C. Littlefield, *Rice and Slaves: Ethnicity and the Slave Trade in Colonial South Carolina* (Baton Rouge: Louisiana State University Press, 1981); Charles Joyner, *Down by the Riverside: A South Carolina Slave Community* (Urbana: University of Illinois Press, 1984); William Dusinberre, *Them Dark Days: Slavery in the American Rice Swamps* (New York: Oxford University Press, 1996); Judith Carney, *Black Rice: The African Origins of Rice Cultivation in the Americas* (Cambridge, MA: Harvard University Press, 2001); Lewis Cecil Gray, *History of Agriculture in the Southern United States to 1860*, 2 vols. (Washington, D.C.: Carnegie Institution of Washington, 1933; reprint, Gloucester, MA: Peter Smith, 1958); Peter A. Coclanis, *The Shadow of a Dream: Economic Life and Death in the South Carolina Low Country, 1670–1920* (New York: Oxford University Press, 1989); Sam B. Hilliard, "The Tidewater Rice Plantation: An Ingenious Adaptation to Nature" in *Geoscience and Man* 12, ed. H.J. Walker (Baton Rouge: Louisiana State Press, 1975): 57–66; and Richard D. Porcher, "Rice Culture in South Carolina: A Brief History, the Role of the Huguenots, and the Preservation of Its Legacy," *Transactions of the Huguenot Society of South Carolina* 92 (1987): 1–22; Mart Stewart, *"What Nature Suffers to Groe": Life, Labor, and Landscape on the Georgia Coast, 1680–1920* (Athens: University of Georgia Press, 1996); Richard D. Porcher and William Robert Judd, *The Market Preparation of Carolina Rice* (Columbia: University of South Carolina Press, 2014).

[2] Edited volumes discussing antebellum tidal rice culture are J.H. Easterby, ed., *The South Carolina Rice Plantation, as Revealed in the Papers of Robert F. W. Allston* (Chicago: University of Chicago Press, 1945); James M. Clifton, ed., *Life and Labor on Argyle Island: Letters and Documents of a Savannah River Rice Plantation, 1833–1867* (Savannah, GA: Beehive Press, 1978); Buddy Sullivan, ed., *Darien Journal of John Girardeau Legare, Ricegrower* (Athens: University of Georgia Press, 2010). For postbellum volumes, see Arney R. Childs, ed., *Rice Planter and Sportsman: The Recollections of J. Motte Alston, 1821–1909* (Columbia: University of South Carolina Press, 1953); Patience Pennington, *Woman Rice Planter* (New York: The Macmillan Company, 1914); Margaret Belser Hollis and Allen H. Stokes, eds., *Twilight on the South Carolina Rice Fields: Letters*

toward antebellum tidal cultivation, questions of the legacy of inland rice remained limited and overlooked.

By focusing on large-scale inland rice production leading up to the Civil War, this chapter discusses the evolution of the plantations and cultivation strategies that relied on reservoirs to irrigate the crop. Nestled between the Awendaw and Mount Pleasant Scarps, eleven plantations actively practiced inland rice cultivation along the upper Wando River after the Revolution. The original property boundaries of these plantations were products of cadastral surveys laid out for proprietary land distribution. Initially, colonial surveyors laid out boundary lines with little regard to the natural landscape. Cadastral survey boundaries were geometrically rigid lines that traversed watercourses and other physical features. As capital-intensive rice planters identified small-stream floodplains and inland swamps as successful agricultural environments, they began purchasing tracts systematically that contained desirable natural features. Once planters understood how specific topographic features facilitated rice cultivation, they began incorporating neighboring tracts to increase rice acreage and their economies of scale. Inland planters' goals were to create a landholding inclusive of each component critical to growing rice and to assemble a labor force large enough to carry out that plan. Planters sought out expansive floodplains and swamps to construct embanked fields and irrigation canals. They also secured bays and streams that fed their reservoirs, eliminating any possibility of upstream planters cutting off their water supply.[3]

Between 1783 and 1860, the Wando River tracts were transformed from square and rectangular 500- and 1,000-acre parcels into irregular 4,300- to 6,600-acre plantations. By the eve of the Civil War, planters consolidated the eleven plantations into four tracts: Charleywood, Fairlawn, Clayfield, and Wythewood. This consolidation began at the Guerin Creek and Wando River confluence and continued twelve miles northeast to Awendaw Creek. Detailed analysis of these four plantations reveal the variations that inland planters made in their cultivation practices, both influencing and contributing to the tidal practices taking shape by the second quarter of the nineteenth century. To accomplish large scale agricultural practices, these inland planters made efforts to expand their reservoirs, canals, and acreage. With a better understanding of these

from the Heyward Family, 1862–1871 (Columbia: University of South Carolina Press, 2010).

[3] Linda M. Pett-Conklin, "Cadastral Surveying in Colonial South Carolina: A Historical Geography" (Ph.D. diss., Louisiana State University, 1986), 133–137.

specific examples of inland rice production, a picture emerges that provides a counterpoint to tidal rice plantations.[4]

Paying attention to the landscape can help one understand the elaborate restructuring of land and water that was necessary to successfully grow rice. As environmental historian Mart Stewart observes, "none of the low-country crops demanded more rigorous reshaping of the environment than wet-culture rice, especially when planters chose to grow it by the tidal flow method."[5] Stewart explains that tidal culture required a critical understanding of hydrology coupled with a sizable labor force to construct miles of embankments and canals, ditches and drains. This chapter will explain that large-scale inland plantations also executed these concepts of water management and cultivation cycles. At the same time, planters had to adjust their agricultural strategies to grow rice in response to shifting ecosystems.[6]

In the Wando River watershed, a twenty-mile finger connecting to the Charleston Harbor, inland rice cultivation resembled tidal practices after the Revolutionary War. Colonial-era inland rice planters adopted concepts of water management – seen in floodgates, trunks, canals, and ditches – similar to tidal rice plantations. Charleywood, Fairlawn, Clayfield, and Wythewood planters continued to incorporate large-scale cultivation strategies during two separate periods after the Revolution to mimic the tidal rice aesthetic of "a huge hydraulic machine." To harness the energy found in tidal rivers' ebb and flow, inland planters had to capture downward flowing water in embanked reservoirs.[7] The first period of post-Revolutionary expansion took place from 1783 to 1839. This stage involved planters who came from families already established in the area prior to the Revolution. These descendants of inland planters

[4] "A Plan of Charleywood Plantation," March 1788, no. 954, John McCrady Plat Collection, Charleston County Register of Mesne Conveyance (CCRMC), Charleston, SC; "Return of Peter Taylor's Estate," 1785, Taylor Family Papers, South Caroliniana Library, University of South Carolina (USC); Ralph H. Willoughby and W.R. Doar, III, "Solution to the 'Two-Talbot' Problem of Maritime Pleistocene Terraces in South Carolina," poster (Columbia: South Carolina Department of Natural Resources, Geological Survey, 2006). Plantations cultivating rice also occurred specifically on the Christ Church Parish side of the Wando River and within Wambaw Swamp, northeast of Awendaw Creek.

[5] Stewart, "*What Nature Suffers to Groe,*" 90.

[6] In 1860, William Dusinberre defines a rice planter as owning at least thirty-five slaves and producing over 100,000 pounds of rough rice, and a large-scale rice planter as owning over 100 slaves; Dusinberre, *Them Dark Days*, 460–461.

[7] Quote from Stewart, "*What Nature Suffers to Groe,*" 98.

made adjustments to sowing techniques and flood schedules to increase their rice yields.[8] The second period took place from 1840 to the eve of the Civil War. This stage of plantation ownership was characterized by a new generation of planter entrepreneurs who sought to capitalize on their prior economic success and diversify their plantation holdings.

Nineteenth-century inland rice cultivation varied between the upper Wando and other watersheds because of the difference in topography and availability of water, as the upper Wando River provided expansive wetlands for rice cultivators. Planters at Charleywood, Fairlawn, Clayfield, and Wythewood increased acreage in an effort to control water for the demands of successful agriculture. Planters' acquisition of specific tracts revealed how they sought to improve their landholdings, as the plantation boundaries enclosed natural features such as waterways, low-lying bays, and wetlands that critically supported inland rice culture. In doing so, these planters managed water flow from impounded reservoirs down to rice fields over a slight grade that moved water over the course of several miles. Fairlawn slaves effectively directed water on and off the fields over a mile and a half, where Wythewood slaves managed a watercourse for more than four miles.[9]

This chapter argues that the success of these four plantations began with the broad Wando River floodplain. The Wando topography enabled planters to position sprawling rice fields on the alluvial marsh and swampland. The Wando River cut through the Lower Coastal Plain forming a serpentine shape from the Awendaw Scarp to the Charleston Harbor. The Wando flows southwest and merges with the Cooper River to form the northeastern half of the Charleston Harbor. The watershed consists of 40,005 acres, or 62.51 square miles above the confluence of the Wando River and Guerin Creek. A ridge extending seaward from the Awendaw Scarp divides the Wando watershed from the Awendaw Creek watershed, which drains into Bulls Bay.[10] Lying no more than seven miles from the Atlantic Ocean, the Wando River watershed consists

[8] Clayfield is also called Clay Fields and Wythewood is also spelled Withewood, Withywood, or Witheywood.

[9] [John D. Legaré] "Account of an Agricultural Excursion Undertaken by the Editor, in the Spring of 1832," *Southern Agriculturalist and Register of Rural Affairs* 5 (July 1832): 354.

[10] Donald J. Colquhoun, *Terrace Sediment Complexes in Central South Carolina* (Columbia: University of South Carolina, 1965), 31–34. Watershed drainage areas are derived from Jennie P. Eidson et al., *Development of a 10- and 12-Digit Hydrologic Unit Code Numbering System for South Carolina, 2005* (Washington, D.C.: United States Department of Agriculture, 2005), 28, 34.

of Pleistocene and Holocene geological deposits.[11] Pleistocene deposits formed barrier islands similar to the present sea islands, as the sea level rose and fell in relation to advances and retreats of the polar ice cap and glaciers. With the ocean shoreline retreating and creating new lines of marine and soil deposits away from the mainland, the former barrier islands became barrier complexes and inland scarps paralleling the current shoreline. The Pamlico Terrace and Awendaw Scarp are products of the Pleistocene. The scarp consists of a mixture of highland well-drained sandy soils with moderately well-drained sandy and loamy soils. These permeable soil types became important foundations for plantation settlements and overland transportation routes, as the firm ground provided hospitable conditions for living and travel.[12]

The 100,000-year-old Princess Anne Terrace lay on the seaward side of the Awendaw Scarp, providing the foundation for the Wando River watershed. The terrace currently begins at approximately twenty feet in elevation and slopes downward from the Santee River Delta toward the Charleston Harbor. The terrace soil is primarily muddy sand, clay, shell, and sand. The Wappetaw Swamp, also called I'on Swamp, lies in the middle of the terrace. The Wappetaw Swamp is a Holocene deposit of Santee clay loam surrounded by Meggett clay loam. The Princess Anne Formation is a result of marine and soil deposits forming behind a former island chain, also called a back-barrier soil formation, of clay and sand when the ocean coastline existed inland from the present shoreline. The back-barrier deposits provided the low-permeable foundations for the Wando floodplain, and also served as a hub for the brickmaking industry

[11] Lucy McCartan, E.M. Lemon, Jr., and R.E. Weems, *Geologic Map of the Area between Charleston and Orangeburg, South Carolina*, Map I-1472 (Washington, D.C.: United States Geological Survey, 1984); Willoughby and Doar, "Solution to the 'Two-Talbot' Problem"; Russell Austin Willis, "Genetic Stratigraphy and Geochronology of Last Interglacial Shorelines on the Central Coast of South Carolina" (M.S. thesis, Louisiana State University, 2006).

[12] Ralph Bailey, Jr., Andrew Agha, and Charles F. Philips, Jr., *Cultural Resources Survey of Part One of the Country Road Environmental Analysis, Francis Marion National Forest, Berkeley and Charleston Counties, South Carolina*, Francis Marion and Sumter National Forests Cultural Resource Management Report 05-01 (Columbia: United States Department of Agriculture, Forest Service, 2008), 9; Colquhoun, *Terrace Sediment Complexes*, 34; Willis, "Genetic Stratigraphy and Geochronology," 1, 8–11; D.R. Soller and H.H. Mills, "Surficial Geology and Geomorphology," in *Geology of the Carolinas: Carolina Geological Society Fiftieth Anniversary Volume*, eds. J.W. Horton and V.A. Zullo (Knoxville: University of Tennessee Press, 1991), 299–301.

on the lower Wando River between the colonial and antebellum eras.[13] Furthest from the ocean, the landward portion of the terrace lacked the stratigraphy of "emergent barrier systems" consisting of sand and shell because the "numerous tidal creeks such as Awendaw Creek and the Wando River" had "dissected," or eroded, the former barrier islands. This development provided desirable soil conditions for the inland planters. The soil content from a sandy loam to dense clay within the "gently inclined slope" of the Princess Anne Terrace played out in how area rice cultivators dealt with water control and management.[14]

The brackish water of the Wando River prevented planters from using tidal irrigation along the floodplains. Water salinity forced planters to choose which rivers would successfully produce rice as the ocean's tidal force pushed salt water into the Lowcountry rivers. Rivers with a small watershed or water located near the ocean produced brackish water, a mixture of salt and fresh water. This water quality killed rice crops growing along the shore. Rivers such as the Wando, with no fresh water from the piedmont and the mountains, could not generate enough force to push the ocean's encroaching waters near the coast. Therefore, plantations situated near the mouth of these rivers could not successfully grow rice by tidal culture compared to plantations located next to rivers, such as the Santee and the Pee Dee, that have larger watersheds and more powerful flows.[15]

Despite the hydrological limitations of the Wando River floodplain, the development of these plantations reflects the surge of optimism experienced after the Revolution. Unlike other inland communities, such as Middle St. John's Parish in Berkeley County, which abandoned rice culture in lieu of cotton farming, upper Wando River planters expanded their rice output. Consistent with historian Joyce Chaplin's interpretation that a "post-Revolutionary scramble to repair property and restore agricultural production" occurred in the South Carolina Lowcountry,

[13] Willis, "Genetic Stratigraphy and Geochronology," 10, 52; Willoughby and Doar, "Solution to the 'Two-Talbot' Problem"; Gene Waddell, *Indians in the South Carolina Lowcountry, 1562–1751* (Columbia: University of South Carolina, 1980), 334–335. For brickmaking on the Wando River, see Lucy B. Wayne, "'Burning Brick and Making a Large Fortune at It Too': Landscape Archaeology and Lowcountry Brickmaking," in *Carolina's Historical Landscapes: Archaeological Perspectives*, eds. Linda F. Stine et al. (Knoxville: University of Tennessee Press, 1997), 97–111.

[14] Willis, "Genetic Stratigraphy and Geochronology," 52; Willoughby and Doar, "Solution to the 'Two-Talbot' Problem."

[15] Hilliard, "Tidewater Rice Plantation," 62, 64–65; Albert Defant, *Ebb and Flow: The Tides of Earth, Air, and Water* (Ann Arbor: University of Michigan Press, 1958), 12.

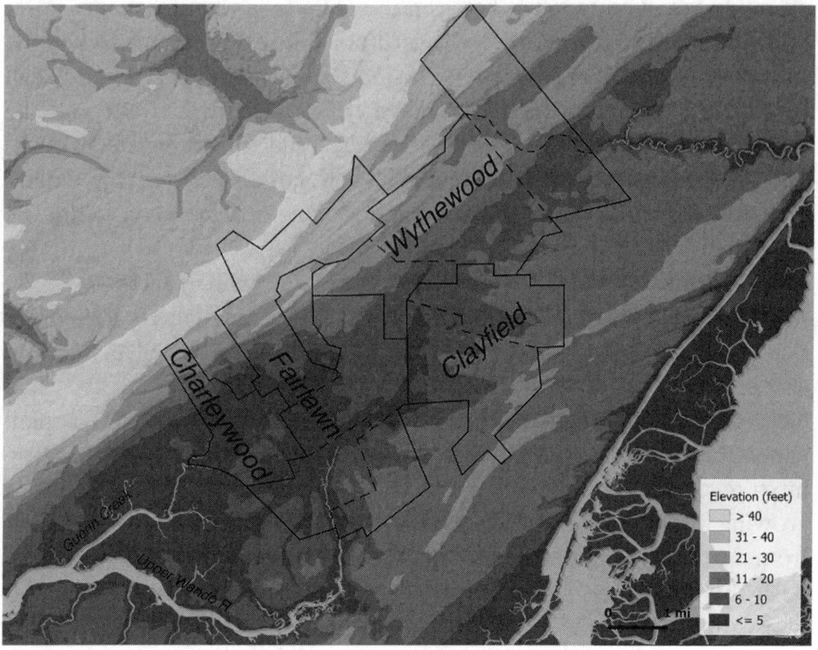

FIGURE 4.1 Upper Wando River rice plantations, ca.1860.
Image adapted by the author. South Carolina Department of Natural Resources, Geographic Information System (GIS) Division

Wando River planters looked at their landholdings with optimism. Planter Major Pierce Butler observed, "ranks of men [thought] of little else than repairing their losses" after the Revolution. However, the "fever of optimism and speculation," according to Marvin Zahniser, "caught up" with several of these planters. As this chapter explains while some ambitious planters became wealthy, others found their economic independence threatened when they fell into debt. In an era when tidal rice cultivation dominated planters' land, labor, and interests, inland rice cultivation continued to exist on the periphery. Wando planters sought out new methods to cultivate the land and control the water through scientific management. When performed successfully, such cultivation techniques rewarded these inland rice planters with bountiful harvests on par with their tidal counterparts.[16]

[16] Chaplin, *An Anxious Pursuit*, 2. Marvin Zahniser, *Charles Cotesworth Pinckney: Founding Father* (Chapel Hill: University of North Carolina Press, 1967), 76; also see

CHARLEYWOOD

Charles Cotesworth Pinckney and Edward Rutledge exhibited this post-Revolutionary optimism through their inland plantation ventures along the Wando and Cooper rivers. The two law partners achieved social status through similar means prior to the Revolution. Both men were educated in England and studied law at the Middle Temple, one of four Inns of Court in London. Once they were practicing law in colonial South Carolina, both Pinckney and Rutledge married daughters of Henry Middleton and Mary Williams and "thus became allied to one of the wealthiest and most political families of South Carolina." Pinckney and Rutledge each became active participants in the political and military campaigns during the Revolution. Pinckney rose to brigadier general and later served as one of the state's delegates to the federal Constitutional Convention. Rutledge was a signer of the Declaration of Independence and served as a delegate of the Second Continental Congress. Rutledge, like Pinckney, took an active military role, fighting in local campaigns until captured after the fall of Charleston in 1780. While serving on the Privy Council in 1782, Rutledge drafted a bill ordering the confiscation of loyalist property. After the war, Pinckney and Rutledge acted upon the bill by co-investing in two confiscated inland plantations, Tippicutlaw and Charleywood, which totaled 3,569 acres.[17]

Pinckney's and Rutledge's optimism was like a facade that obscured Charleywood's economic problems. "The future looked bright to Pinckney," according to his biographer, "so bright that he failed to calculate carefully the consequences of heavy indebtedness." In the first year of ownership, Pinckney and Rutledge lost one-third of their crop to drought. They were constantly behind in payment. Peter Taylor, who sold Charleywood to Pinckney and Rutledge, wrote to his attorney Rodger Smith that he had not received the 1788 payment of £450 sterling. Having moved a labor force from a Middleton family plantation to Charleywood, Rutledge scrambled to plant 200 acres of rice for payment. Even after Taylor's death in 1789, Pinckney and Rutledge still owed his widow a significant

Jerome J. Nadelshaft, *The Disorders of War: The Revolution in South Carolina* (Orono: University of Maine Press, 1981), ch. 11: "No Reason to Despair," 191–217.

[17] N. Louise Bailey and Walter Edgar, eds., *Biographical Directory of the South Carolina House of Representatives*, vol. II: *The Commons House of Assembly, 1692–1775* (Columbia: University of South Carolina Press, 1977), 525–527, 573–574, quote: 573; Richard Clow, "Edward Rutledge of South Carolina" (Ph.D. diss., University of Georgia, 1976) 204–205, 321; Zahniser, *Charles Cotesworth Pinckney*, 76.

balance. Pinckney expressed a "great uneasiness" to Rutledge for their outstanding debt in 1797, after realizing their $2,600 (£586 sterling) payment from a sale of 190 barrels did little to diminish their delinquency to the Taylor estate. After Rutledge died in 1800, Pinckney sold Charleywood to Thomas Wigfall at public auction for 4,050 guineas (£4,253 sterling). Wigfall, in turn, incorporated Charleywood into his 1,708-acre Bull Head plantation.[18]

FAIRLAWN

Fairlawn Plantation adjoined Charleywood's eastern boundary and provided a stark contrast in size and infrastructure. It belonged to the Rose family, and by 1845, on the death of Hugh Rose, the Fairlawn Plantation was almost 7,000 acres. John Rose, Hugh's father, initiated the Richfield plantation, the forerunner of Fairlawn. He had become one of the largest shipwrights in the British colonies building both military and commercial fleets. Despite this success, the elder Rose sold his share of the Wando River boatyard and reinvested his capital into rice plantations. Rose's decision is a telling tale of the grain's profitability and desire. However, John Rose sided with the crown during the Revolution and, like Taylor and fellow loyalists, lost his property to the new republic. In 1783, Hugh Rose paid a large fine of £49,200 sterling to reclaim the family seat, the confiscated Richfield plantation. The Rose family captured water from Wappetaw Swamp to flood the Richfield rice fields on the Wando river floodplain. In 1794, Hugh Rose paid Thomas Screven £4,300 sterling for neighboring Fairlawn and consolidated the two plantations to form a 4,462-acre tract. Capturing water from the Pamlico Terrace bays, Screven's slaves carved out more than 200 acres of rice fields that separated Charleywood from Richfield. Fairlawn used water control characteristics

[18] Zahniser, *Charles Cotesworth Pinckney*, 76; Rodger Smith to Peter Taylor, 21 August 1786, Taylor Family Papers, USC; Peter Taylor to "My Dear Friend" [Rodger Smith], 26 November 1788, Taylor Family Papers, USC; John LaFayette Brittain, "Two Recently Discovered Letters of Charles Cotesworth Pinckney: Another Glimpse into the Mind of an Eighteenth Century Man of Affairs," *South Carolina Historical Magazine* 76 (January 1975): 17; "Charles Cotesworth Pinckney to Thomas Wigfall, release," 15 February 1802, Deed Book H7: 330, CCRMC; "John Wigfall Will," Miscellaneous Inventories and Wills, Charleston County, Will Book C: 44, South Carolina Department of Archives and History (SCDAH), Columbia, SC. Historic exchange rates converted through the Economic History Association's website: www.measuringworth.com/calculators/exchange/ (accessed May 1, 209). (1797) £1 = $4.44; $2,600 / 4.44 = £585.585. (1800) 1 g. = £1.05; 4,050 x £1.05 = £4,252.5.

that were similar to Charleywood's to flow water in a linear path from the scarp to the floodplain. The plantation relied on water flow from the Awendaw Scarp and from Guerin Creek, providing several directions in which Rose could control field irrigation. During the nineteenth century, Rose added two neighboring tracts to the south. He also bought two neighboring tracts to the north, securing water control from the Awendaw Scarp to feed Fairlawn's reservoirs. By incorporating the additional tracts during the first two decades of the nineteenth century, Rose initiated an inclusive network of canals, embankments, and fields that made him a prominent inland rice cultivator.[19]

Rose continued to expand his plantation boundaries toward the Wando River by purchasing Windsor Forest from Daniel Ward in 1797. The 590-acre tract extended Rose's rice fields and drainage canals south of Fairlawn to the upper Wando River. By annexing this property, Rose not only acquired 100 acres of "prime swamp land" for rice cultivation, but he also secured better drainage and navigability at Ward's, or Wappetaw, Bridge. Ward built a "landing place" on the creek, connected by a road over "high land" to his plantation and slave settlements less than one-half mile from the landing, allowing for efficient transportation of commodities from field to market.[20]

[19] John Rose documented in: Douglas Catterall, "The Worlds of John Rose: A Northeastern Scot's Career in the British Atlantic World, c.1740–1800," in Angela McCarthy, ed., *A Global Clan: Scottish Migrant Networks and Identities since the Eighteenth Century* (New York: Tauris Academic Studies, 2006), 67–94; "Hugh Rose to William Parker and Edward Blake, Commissioners of the Treasury of South Carolina, Mortgage" 17 December 1783, Deed Book T5: 453–454, CCRMC; "Thomas Screven and His Wife Amaranthia to Hugh Rose, Release" February 5, 1794, Deed Book K6: 290, CCRMC; "William J. Grayson to James Rose, Mortgage" 12 April 1853, Deed Book Y12: 374, CCRMC; "Plan of Fairlawn Plantation in St. Thomas Parish, Charleston District," May 1794, no. 4339, John McCrady Plat Collection, CCRMC.

[20] John Rose documented in: Douglas Catterall, "The Worlds of John Rose: A Northeastern Scot's Career in the British Atlantic World, c.1740–1800," in Angela McCarthy, ed., *A Global Clan: Scottish Migrant Networks and Identities since the Eighteenth Century* (New York: Tauris Academic Studies, 2006), 67–94; "Hugh Rose to William Parker and Edward Blake, Commissioners of the Treasury of South Carolina, Mortgage" 17 December 1783, Deed Book T5: 453–454, CCRMC; "Thomas Screven and His Wife Amaranthia to Hugh Rose, Release" February 5, 1794, Deed Book K6: 290, CCRMC; "William J. Grayson to James Rose, Mortgage" 12 April 1853, Deed Book Y12: 374, CCRMC; "Plan of Fairlawn Plantation in St. Thomas Parish, Charleston District," May 1794, no. 4339, John McCrady Plat Collection, CCRMC; "Plan of Land on Wapetaw Creek," May 1794, no. 4362, John McCrady Plat Collection, CCRMC. Wappau Creek and Wapetaw Creek are variations of Wappetaw Creek; see: Waddell, *Indians in the South Carolina Lowcountry*, 333.

By the turn of the century, Fairlawn began to take the shape of a large-scale rice plantation with 865 acres devoted to embanked fields, drains, and canals. In April 1802, Rose paid £806 12s to Thomas Wigfall for eighty acres of Charleywood rice fields bordering Fairlawn. The purchase served two purposes. The first, obviously, was to expand Fairlawn's rice lands. During Taylor's ownership, the Charleywood parcel only had half the eighty-acre tract cultivated with the lower half consisting of swampland. Between 1794 and 1800, however, Pinckney and Rutledge expanded the field system to enclose the entire quadrant. The tract of quartered and drained fields provided Rose with an additional parcel of reliable rice land. The Pleistocene back barrier provided a desirable environment for constructing level rice fields on Santee clay loam, yet offered enough slope from the scarps to the river for enslaved cultivators to initiate proper drainage.[21] Efficient drainage through Fairlawn was the second reason for Rose's interest in the Wigfall property. The square tract jutted into Fairlawn, and prevented continuous drainage from the western Fairlawn fields. Water draining from the acres had to circumnavigate the Charleywood tract. By incorporating the eighty acres into the Fairlawn rice fields, Rose constructed a canal traversing the newly acquired rice fields and connected the drainage canal. Because of the new canal, water traveled in a straight direction flowing from the upper to the lower rice fields along the western Fairlawn boundary.[22]

Rose continued acquiring property bordering Wappetaw Swamp that formed a plantation holding spanning the Awendaw and the Mt. Pleasant Scarps. Hugh Rose completed this expansion in 1807 by purchasing Capers Plantation for £3,000 SC. The 765-acre tract connected to Fairlawn's southeast boundary and to the east of Windsor Forest. Separating Fairlawn from the Caper's tract was a drainage canal originating in Rose's property to drain water from Mayrant's Reserve and control water in the Wappetaw Swamp. Rose's annex allowed his enslaved cultivators to further tap into the Wappetaw wetland, extending embanked rice fields by 165 acres. Under the Capers family, water flowed through Fairlawn's undeveloped wetlands and from Durant Plantation's discharged water

[21] "Thomas Wigfall to Hugh Rose, Release," 17 February 1802, Deed Book G7: 393, CCRMC; Willoughby and Doar, "Solution to the 'Two-Talbot' Problem"; "A Plan of Charleywood Plantation," March 1788, CCRMC.

[22] "Thomas Wigfall to Hugh Rose, Release," 17 February 1802, Deed Book G7: 393, CCRMC; Willoughby and Doar, "Solution to the 'Two-Talbot' Problem"; "A Plan of Charleywood Plantation," March 1788, CCRMC; "Plan of Fairlawn Plantation," May 1794, CCRMC.

from the east. Similar to Fairlawn, Capers Plantation used central and flanking canals to move water on, around, and off the rice fields. With a central infrastructure in place, Rose expanded upon this landscape with more enslaved labor to form an inclusive field system settled into the Wappetaw Swamp.[23]

CLAYFIELD

Clayfield Plantation followed a similar path to Fairlawn in terms of size and productivity. Jacob B. I'on Sr.'s plantation sat upstream from Fairlawn and Capers on the Wappetaw Swamp. Between 1765 and his death in 1796, I'on systematically assembled five tracts of land totaling 2,350 acres and enslaving 122 people at Clayfield. I'on's father, Capt. Richard I'on, initiated the family's involvement in rice culture by purchasing the 500-acre "Swamp Plantation" from George Bennison's estate in 1748. The tract consisted of a mixture of Holocene swampy lowlands and Pleistocene swells of sandy pinelands. Clayfield rice cultivation took place in the Wappetaw Swamp, a desirable location to grow rice because of the high concentration of fertile decaying organic matter and high volume of water.[24] After 1765, Jacob B. I'on, Sr. purchased the central part of Clayfield Plantation from John Bruce's estate. The 400-acre tract consisted of a peninsula of high ground surrounded by the rice fields in Wappetaw Swamp. The Bruce settlement, outbuildings, and slave settlement clustered together on the narrow highland finger.[25]

I'on further expanded his Clayfield property by purchasing a fifty-acre triangle abutting Hampton Plantation to the north. This tract, purchased from John Rose in 1778, provided more wetlands for I'on to create the

[23] "Estate of Gabriel Capers to Hugh Rose, Release," 1 April 1807, Deed Book U7: 241, CCRMC; £3000 SC is South Carolina pounds; "Plan of Fairlawn Plantation," May 1794, CCRMC; "Plan of a Plantation or Tract of Land in the Parish of Christ Church, Charleston District," February 1807, no. 6064, John McCrady Plat Collection, CCRMC.

[24] Karen G. Wood, *Site Evaluation on Three Sites at Historic Clayfield Plantation, Wambaw Ranger District, Francis Marion National Forest, South Carolina,* Francis Marion and Sumter National Forests Cultural Resource Management Report 91-30 (Columbia: United States Department of Agriculture, Forest Service, 1991), 15; Willoughby and Doar, "Solution to the 'Two-Talbot' Problem;" N. Louise Bailey and Walter Edgar, eds., *Biographical Directory of the South Carolina House of Representatives*, vol. III: *1775–1790* (Columbia: University of South Carolina Press, 1981), 368–369.

[25] Wood, *Site Evaluation on Three Sites*, 15–17, 27–48; H. Roy Merrens and George D. Terry, "Dying in Paradise: Malaria, Mortality, and the Perceptual Environment in Colonial South Carolina," *Journal of Southern History* 50 (November 1984): 543–544.

third reservoir. With this addition, Clayfield used two reservoirs flooding the upper fields and one reservoir flooding the lower fields. Water from all three of these reservoirs flowed west into the fields, yet the hydrology of Wappetaw Swamp let the water from the upper reservoirs work its way around the high-ground settlement and through the rice fields downstream. The Revolutionary War led to a twelve-year hiatus of I'on property expansion.[26]

Jacob B. I'on, Sr. began his post-war expansion purchasing 400 acres of Hampton Plantation from Joseph Wigfall in 1790.[27] Wigfall was the middle of three surviving sons of Capt. Samuel and Catherine Wigfall of Willow Hall and uncle to Thomas Wigfall of Charleywood. As discussed in Chapter 2, the Wigfall family had been raising livestock and growing rice in the area since 1702, resulting in generations inheriting or purchasing property throughout the Princess Anne Terrace. Hampton Plantation's topography resembled Clayfield's with a mixture of high sandy pinelands and low-lying wetland bays traversing the property. The tract that Wigfall sold to I'on was not prime rice land because of limited topography to manage water in impounded fields. The tract, however, enabled I'on to expand his lower reservoir, harvest timber, and cultivate high ground.[28]

With the upper Wando River providing a boundary line, I'on accumulated land toward the Mt. Pleasant Scarp opposite of Rose's plantation continuing up to the Awendaw Scarp. I'on began to acquire the fifth tract in 1796 that would secure the southern half of Wappetaw Swamp. The property consisted of 1,000 acres along the Clayfield border. Originally part of a 1,700-acre proprietary grant to John White in 1705, the tract

[26] Wood, *Site Evaluation on Three Sites*, 15, 20; US Geological Survey, Sewee Bay Quadrangle, South Carolina [map], revised 1959, 1:24,000, 7.5 Minute Series (Reston, VA: US Department of the Interior, USGS, 1969).

[27] I'on's Hampton Plantation is not the plantation of the same name located on the South Santee River, made famous by Archibald Rutledge, and now a state park.

[28] Henry A.M. Smith, "The Baronies of South Carolina: The Seewee Barony," *South Carolina Historical and Genealogical Magazine* 12 (July 1911): 110–113; Robert M. Weir, *Colonial South Carolina: A History* (New York: KTO Press, 1983; reprint, Columbia: University of South Carolina Press, 1997), 80–82; G. Ishmael Williams et al., *An Archaeological Survey of 3,438 Acres in the Coastal Area, Wambaw and Witherbee Districts, Francis Marion National Forest*, Francis Marion and Sumter National Forests Cultural Resource Management Report 91-45 (Columbia: United States Department of Agriculture, Forest Service, 1993), 46–48; "A Plan of Thomas Barkesdale's Plantation on Wappetaw," 1792, no. 5972, John McCrady Plat Collection, CCRMC; Alston Deas, "Wigfall of Christ Church Parish," Wigfall Family Papers, College of Charleston Special Collections, Charleston, SC, 5–6.

passed through several hands until it was inherited by Susannah Durand, a minor. In 1796, Susannah's husband Henry Durand executed a promissory title to I'on once Susannah reached twenty-one, but I'on died that year. After settling legal matters with the Durant family, Col. Jacob Bond I'on Jr. acquired the tract in 1815. The tract expanded the inland field infrastructure further south, following the contours of Wappetaw Swamp, and drained into Capers Plantation. With the Durant tract, Clayfield acreage tripled with a total of 500 acres devoted to rice and 289 acres in reserve water.[29]

WYTHEWOOD

Wythewood Plantation, located on the headwaters of the Wando River, provided a model of creative inland rice irrigation and drainage. Robert Quash Jr. consolidated 1,686 acres by 1789, naming the plantation after the Bristol, England district. Robert's father, Robert Quash, Sr., made his fortune at Fishbrook Plantation, on Turkey Creek. Turkey Creek and Nicholson Creek form Huger Creek and the headwaters of the East Branch of the Cooper River. Both Fishbrook and neighboring Windsor Plantation represented the extensive inland rice cultivation in this watershed by the mid-nineteenth century. Upon his father's death in 1772, Robert Quash, Jr. inherited Fishbrook and Cypress Pond, a 1,004-acre Hasell family plantation. The younger Quash sold the Hasell property and converted the capital to help purchase four tracts on the headwaters of the Wando between 1770 and 1790. While continuing to manage Fishbrook, which he also expanded by an additional 600 acres, Robert began assembling strategic tracts to make up the core area of Wythewood.[30]

Robert grew up learning the methods of inland rice cultivation on Fishbrook. He continued to practice cultivation methods that his father successfully applied to the land. Taking advantage of his father's inheritance and £20,000 in bonds, he captured the speculative spirit by many of the wealthy new-republic citizens. With Robert's mother living at

[29] Wood, *Site Evaluation on Three Sites*, 16, 21; Sewee Bay Quadrangle, 1969; "Plan of a Plantation Called Clayfield in Christ Church Parish," May 1816, no. 4287, John McCrady Plat Collection, CCRMC.

[30] "Robert Quash Will," Miscellaneous Inventories and Wills, Charleston County, Will Book 14:205, SCDAH; "G.A. Trenholm to F.D. Quash, Mortgage," 15 February 1853, Deed Book B13: 89, CCRMC; "G.A. Trenholm to William Lucas, Release," 13 February 1860, Deed Book J14: 51, CCRMC.

Fishbrook, Quash systematically bought tracts lying between the Wando headwaters and the western tributary of Awendaw Creek. Two tracts were 500-acre parcels granted by the Lords Proprietors in 1708 and 1710. A third tract of 643 acres was part of a 48,000-acre grant to landgrave John Bayley in 1698. Robert also purchased unclaimed acreage in June 1788 to complete Wythewood's central portion. The three tracts did not produce a significant output of rice until Quash consolidated these tracts into a 1,640-acre plantation.[31]

By 1790, Robert took the lessons of water control gained while living at Fishbrook and implemented a similar system at Wythewood. Like his father's Fishbrook slaves, his Wythewood slaves carved out a series of field divisions that followed the natural watercourse. Flanking canals identical to Fishbrook bordered the Wythewood rice fields. The younger Quash learned lessons from his ancestors' trial-and-error. While the older Fishbrook fields varied in size and placement, representing Quash's attempt to understand the agricultural system and its relationship with the topography, Robert's enslaved labor force created consistently shaped fields within the floodplain. Eventually, the central Wythewood drainage system would form the basis of the Wythewood Canal, a central artery transferring water to and from rice fields, while also serving as a seven-mile navigable route to the Wando River.[32]

Francis Dallas Quash inherited Wythewood while attending Harvard in 1811. Upon receiving an M.A. from the university in 1817, Quash returned to the Lowcountry to pursue planting. Quash dramatically improved the plantation's infrastructure for better water control by adding additional acreage and connecting the flanking canals to the Wando and Awendaw watercourses. Quash expanded Wythewood acreage by purchasing Cypress Hedge from Elizabeth Bonneau for $6000 in 1819. Cypress Hedge was the Bonneau family seat and originated from a 1709 proprietary land grant to Anthony Bonneau. The 300-acre

[31] Bailey et al., *Cultural Resources Survey of Part One of the Country Road Environmental Analysis*, 78; Williams et al., *Archaeological Survey of 3,438 Acres in the Coastal Area*, 214; "Trenholm to Lucas," 51; M.L. Walker, *Abstract of Title Covering A.C. Lumber Company Tracts #1 and #1-II, Charleston County, South Carolina, Containing 18,530 Acres*, vol. II (Charleston: United States Department of Agriculture, Forest Service, 1934), 81–119.

[32] "A Plat Exhibiting the Site of a Canal, with the Adjacent Country, between Wando, Santee, and Sampit Rivers," 790, Maps and Muniments Series, South Carolina Historical Society (SCHS), Charleston, SC; "Map of Christ Church Parish 1824," State Plat Books, SCDAH.

plantation was a portion of a larger 742-acre tract passed through the Bonneau family until Henry Bonneau's death in 1811. In accordance with Henry's will, the heirs subdivided the plantation with his niece Elizabeth Vanderhorst Bonneau receiving the central Cypress Hedge tract. Quash purchased this 300-acre parcel from Bonneau to expand his rice fields and canal system. By 1824, Quash had 96 enslaved people working at Wythewood and expanded his labor force to 152 people cultivating 400 acres of rice by 1840.[33]

Inland Rice Schedules and Scientific Agriculture

Because of the large acreage devoted to rice cultivation, Wando planters depended upon specific irrigation and drainage strategies that eventually mimicked tidal plantations. By following the evolution of irrigation and drainage systems at Charleywood, Fairlawn, Clayfield, and Wythewood, a picture emerges of how topography intersected with cultivation in inland rice production. *Southern Agriculturalist* editor John D. Legaré rightfully noted the difficulty of growing inland rice, as the upper Wando planters "depended altogether on reservoirs for their supply of water, which of course render[ed] the crops a little precarious." Because inland planters were limited to the amount of water needed to grow rice, each planter modified the basic procedure to cultivate the grain. Legaré observed how "owing to their respective situations, and sometimes to a difference of opinion," no two cultivation practices were the same. Planters had to determine the best methods for impounding water, irrigating the fields, and draining the field divisions.[34]

In the case of Charleywood, challenges stemmed from irrigating broad rice fields with limited access to water. As Thomas Smith's correspondence with Peter Taylor showed, the lack of impounded water from the 1773 drought left many fields short of adequate flooding. Before the Revolution, Charleywood hands flooded 200 acres from a forty-acre reservoir. The 1773 drought revealed limitations in Charleywood's water supply, as Smith noted only 110 acres of rice "stands tolerably" from the lack of reservoir water "and that will suffer if we have not rain soon." To solve this problem, Smith more than doubled the impounded water

[33] Walker, *Abstract of Title Covering A.C. Lumber Company Tracts #1 and #1-II*, 125–138; Bailey and Edgar, *Biographical Directory of the South Carolina House of Representatives*, vol. II: *1320–1321*.

[34] [Legaré] "Account of an Agricultural Excursion," 354.

volume to flood many of the fields below and to the east over the next decade. By 1785, Taylor's plantation had 300 acres of "prime swamp" land with eighty-five slaves toiling in the fields.[35]

In order to secure enough impounded water for adequate flooding, inland planters sacrificed potential rice fields for necessary reservoirs. Transferring sixty acres from rice fields to a reservoir by 1785, Taylor secured more impounded water in return for growing less rice. Twenty-one years after the drought, Pinckney boasted to Ralph Izard that "I find with two-hundred twenty workers you have not made this year more rice than Ed Rutledge and myself have at Charleywood with ninety." In 1795, Rutledge optimistically exclaimed the possibilities of new water diversion methods. He stated that the old overseer passed away, whose position was filled by "Mr. Powell." Powell recommended digging a 100-yard drainage canal, which Pinckney and Rutledge "could command … twice as much water" from the middle reserve to the lower reservoir. Powell also advised Rutledge to dig a second canal from the lower reservoir to the Fairlawn Canal. Besides flooding and draining rice fields, the second canal powered a rice mill that Rutledge believed could "beat twenty barrels a day" and ultimately "beat two thousand barrels" from the reservoir surplus. Rutledge, quite pleased with this position, related to Pinckney that the new canals and mill "fused a new life into our people," presumably from the more efficient market preparation of rice that was previously accomplished by mortar and pestle.[36]

While Pinckney and Rutledge expanded Charleywood's reserve system at the expense of rice acreage, Rose acquired enough property to diversify his Fairlawn watercourses to systematically irrigate three large field divisions. After merging Fairlawn and Richfield, his plantation had impounded water covering 569 acres to flood more than 650 acres of rice fields. Like Charleywood, the upper Fairlawn reservoir tapped into tributaries flowing from the top of the Awendaw Scarp. Unlike

[35] "Plan of Charleywood Plantation," March 1788; Thomas Smith to Peter Taylor, 14 July 1773, Taylor Family Papers, USC; "Return of Peter Taylor's Estate," [1785], Taylor Family Papers, USC; Peter Coclanis, "Thickening Description: William Washington's Queries on Rice," *Agricultural History* 64 (Summer 1990): 10–12. For Legaré's connection to scientific agriculture and larger themes of agricultural reform, see: Theodore Rosengarten, "*The Southern Agriculturalist* in the Age of Reform," in *Intellectual Life in Antebellum Charleston*, eds. Michael O'Brien and David Moltke-Hansen (Knoxville: University of Tennessee Press, 1986), 279–294.

[36] "Two Letters from Charles Cotesworth Pinckney to Ralph Izard," *South Carolina Historical and Genealogical Magazine* 21 (October 1940): 151; Edward Rutledge to Charles Cotesworth Pinckney, 9 June 1795, Huger Family Papers, SCHS.

Charleywood, however, Rose embanked large bays that ran four and a half miles along the base of the scarp. To secure this water control, Rose purchased two tracts totaling 1,729 acres mixed with high pineland and low-lying bays. Three canals drained the massive bay into the first reservoir. From the upper reservoir, water irrigated the upper field systems and also flowed into a second 108-acre reservoir. This second reservoir, called Penny Dam, enabled enslaved trunk minders to flood quickly the adjoining fields situated below the reservoir dam. By containing water in Penny Dam, trunk minders could flood lower divisions more efficiently and accurately. By not having to wait for water to travel approximately 2,000 feet from the upper reservoir located on the northern side of the Public Road, the trunk minders could accurately determine when to shut down water flow without excess water breaching embankments or flowing into adjoining fields.[37]

Adjacent to the brackish surges of the Wando River, the lower division of Fairlawn's fields resembled the layout of tidal counterparts without the use of "estuary hydrology" to irrigate the land. While the canals and embankments at Fairlawn had all the characteristics of a tidal plantation, the lower division of rice fields received water from a sprawling bay paralleling the base of the Awendaw Scarp. The bay flowed into Fairlawn from the north and from the east. In a fashion similar to Fairlawn's upper reservoir, this irrigation system consisted of an upper and lower reservoir connected by a natural watercourse. On the northeastern half, Rose dammed 143 acres to form another "bay swamp" reservoir. The impounded water then flowed into the 239-acre Mayrant's Reserve. Rose's enslaved laborers were forced to build a dam more than a mile long to withhold the water covering this bay.[38]

Critical in connecting water with the fields, canals were the arteries of the inland plantations. Unlike tidal plantations that had fields adjoining the rivers, inland plantations relied on a large amount of water from a variety of sources. At Fairlawn, large canals intersecting smaller ditches and drains connected the four reservoirs to the field systems. Besides

[37] "Plan of Fairlawn Plantation," 1794; Sewee Bay Quadrangle, 1969; US Geological Survey, Ocean Bay quadrangle, South Carolina [map], Revised 1992, 1:24,000, 7.5 Minute Series (Reston, VA: US Department of the Interior, USGS, 1992).

[38] Stewart, "*What Nature Suffers to Groe*," 98–99; "Plan of Fairlawn Plantation," 1794; Sewee Bay Quadrangle, 1969; US Geological Survey, Ocean Bay Quadrangle, South Carolina [map], Revised 1992, 1:24,000, 7.5 Minute Series (Reston, VA: US Department of the Interior, USGS, 1992); Willoughby and Doar, "Solution to the 'Two-Talbot' Problem."

serving as a conduit from the reservoirs to the fields, the fifteen-foot-wide channels also served as flanking canals for Fairlawn's rice field network. In 1794, a central canal ran the distance of the Fairlawn rice fields, one and a quarter mile, from Penny Dam to a lower canal. By the middle of the nineteenth century Rose elongated this central artery to flow the length of Fairlawn, from the upper bay reserve to the Wando, a length of three and a half miles. Two canals drained adjacent fields, with the Fairlawn Canal draining portions of the upper division and the Wappetaw Canal draining the remaining sections. By Rose's death in 1841, Fairlawn consisted of 600 acres devoted to reservoirs and canals supporting 800 acres of rice fields "under bank."[39]

Upper Wando planters actively practiced innovative agricultural techniques and contributed to the advancement of scientific agriculture.[40] By 1825, Rose devoted 5,050 acres to his agricultural pursuits. His expansive property holding sat in both Christ Church and St. Thomas/St. Denis Parishes, and Rose utilized 199 slaves to maintain the plantation's complex infrastructure. Fairlawn's plentiful water reserves and large labor force enabled Rose to maintain sufficient irrigation and cultivation to maintaining his rice crop. While directing his enslaved population to grow the cash crop, Rose devoted time, labor, and acreage to experimenting with new rice cultivation practices.[41] Through trial and error, Rose established a successful routine for cultivating rice in this inland landscape. Rose (or his enslaved driver) instructed slaves to plant rice in

[39] "Plan of Fairlawn Plantation," 1794; Sewee Bay Quadrangle, 1969; US Geological Survey, Ocean Bay Quadrangle, South Carolina [map], Revised 1992, 1:24,000, 7.5 Minute Series (Reston, VA: US Department of the Interior, USGS, 1992); *Charleston Mercury*, 8 January 1845: 3.

[40] See Chapter 7 for a further explanation of scientific agriculture.

[41] "Return of Hugh Rose, Taxable Property in the Parish Christ Church," 1824, SCDAH; "Return of Hugh Rose, Taxable Property in the Parish of St. Thomas and St. Dennis," 1824, SCDAH. Rose regularly published his agricultural experimentations in the *Southern Agriculturalist and Register of Rural Affairs*, associated with the Agricultural Society of South Carolina (Charleston, SC); see: "Queries on the Culture of Rice; by William Washington, with Answers, by Hugh Rose of St. Thomas," *Southern Agriculturalist and Register of Rural Affairs* 1 (April 1828): 166; "On the Cultivation of Clover; by Hugh Rose," *Southern Agriculturalist and Register of Rural Affairs* 1 (June 1828): 249; "Additional Information on the Cultivation of Clover; by Hugh Rose," *Southern Agriculturalist and Register of Rural Affairs* 1 (August 1828): 346; "On Open Rice Planting; by Hugh Rose, Esq.," *Southern Agriculturalist and Register of Rural Affairs* 2 (August 1829): 370; "On the Use of Chloride of Lime as a Preventative against Country Fever; by Hugh Rose," *Southern Agriculturalist and Register of Rural Affairs* 4 (May 1831): 250.

half-acre divisions of 150 feet square. The cultivation season began with Rose's field hands cutting "equi-distant" trenches, fourteen inches apart, in the early spring during the last week in March or the first week of April. Unlike his peers who harrowed their fields once before planting, Rose broke the soil twice to create a "dry state" for his seed. The Wappetaw wetlands were composed primarily of "strong tenacious blue clay" with low water permeability. Rose believed the soil produced an "elongated and weakened" seedling because "its energies [were] nearly destroyed" from the crop's roots trying to penetrate the firm clay loam. To counter this problem, Rose forced his enslaved cultivators to "reduce [the soil] to a good tilth, by ploughing a portion, and always digging what [his slaves] are unable to plough." He believed that preparing the soil by using a hoe to break down the clay loam was the most important part in preparing inland rice cultivation.[42]

By 1826, Rose began experimenting with open, or uncovered, rice seed planting. Inland and tidal planters developed open seed, or open trench, planting in the mid-1820s as an attempt to solve problems of declining rice yields caused by worn-out soil. The cultivation practice involved slaves (usually children) encasing rice seeds with clay and sowing them in uncovered trenches. Trunk minders would then slowly let one and a half foot of water onto the fields, so flowing would not wash away seeds. During the first flooding, or "sprout flow," water eroded the trench banks causing soil to cover the grain. The seeds sat underwater for approximately twenty-one days until the seeds sprouted and germinated through the soil. After this stage, water was gradually drawn off, so as not to damage the delicate crop. Fields dried for fifteen days and enslaved field hands removed any competing weeds and volunteer rice. As the seedlings grew to a height of two to three feet, the trunk minders let on a second flooding, or "stretch flow," for twenty-one days. During this flow,

[42] [Washington and Rose] "Queries on the Culture of Rice," 166; [Rose] "On Open Rice Planting," 370; Columella, "Answers to Queries on the Culture of Rice," *Southern Agriculturalist and Register of Rural Affairs* 6 (May 1833): 225. Nineteenth century geographer Jedidiah Morse commented "Rice ground is prepared only by effectually securing it from the water, except some higher parts of it, which are sometimes dug up with a hoe, or mellowed by a plough or harrow." Morse further noted: "Those who have water in reserve, commonly let it in upon their rice, after first going through with the hoe, while it is young, though it is deemed best to keep out the grass without this aid, by the hoe only." Jedidiah Morse, *The American Universal Geography* (Boston: Thomas & Andrews, 1802), 687.

floodwaters would lift up the "trash" of pulled weeds and stalks. A second and possibly third hoeing took place during the forty-day period after trunk minders let the water off the fields. Finally, the harvest flow took place until the rice crop reached maturation. The harvest flow required the most amount of water because the flooding needed to be as high as the plants, to support the heavy panicle that was sprouting from the stalk.[43]

To the "adventuresome, experienced, and scientific-minded planters," according to Albert House, open-seed cultivation was rewarding in three ways: planters used less labor because slaves only had to hoe three instead of five times, rice yields increased from the clay adding nutrients, and less flooding shortened the growing season. However, open planting was "a gamble" because the rice crop was more susceptible to natural disasters. Wind or rainstorms could eradicate the plant during the early weeks, usually before the second flooding, because the shallow root systems could not anchor the plant compared to rice seeds covered in soil. Rose devoted a percentage of his crop to the practice for three years before declaring that, with "experience and attentive observation" requiring "vigilance and daily inspection," open rice planting was superior to other practices. He recognized that the practice created a more consistent growth that "ripens more uniformly" with less overall time and labor devoted to growing rice. To achieve this result, Rose's cultivators pre-soaked seeds twelve to twenty-four hours and then dried them the night before sowing on the barn floor. Rose ordered that two bushels of seeds per acre – one-quarter to one-half bushel an acre less than his peers' amount – planted in one day with the fields flooded the same night to protect the seed from rice birds. Rose strategically planted late in the season, after mid-April when Bobolinks had finished their migration north and there was less chance of frost killing the crop. Depending on pre-existing moisture content in the soil, Fairlawn trunk minders would only leave the first flooding on the fields from five to eight days, so the

[43] [Rose] "On Open Rice Planting," 370; Albert V. House, Jr., *Planter Management and Capitalism in Ante-bellum Georgia* (New York: Columbia University Press, 1954), 33–34; Albert V. House, Jr., "Charles Manigault's Essay on the Open Planting of Rice," *Agricultural History* 16 (October 1942): 184–186; Clifton, *Life and Labor*, xxii–xxiii, 101–108; Dusinberre, *Them Dark Days*, 189; A.M.H. "Rice Cultivation," *Southern Cultivator* 24 (December 1866): 278.

crop would not become weak or rot. This practice shortened the initial flood stage by at least a week compared to the closed seed method.[44]

Rose's attentiveness to field conditions represented how planters paid attention to changing weather patterns, and in turn, altered their cultivation strategies to work within the natural world. Before Rose exclusively adopted the open seed planting, his enslaved trunk-minders flooded fields only when the seeds required water. The moist Fairlawn landscape required less water than inland fields on more permeable soil. Even with the plentiful water at hand, Rose still cautiously used his reserves. He admitted that his irrigation strategies resulted from "being dependent on reserve water." Rose did not instigate "point flowing," the third flood of the rice cultivation cycle. Instead, Rose advocated against early or long watering on the fields, believing that too much moisture led to an inferior crop. Jordan Myrick, a contemporary in the Agricultural Society of South Carolina, also went against common tidal practice by eliminating a "point flow" on his fields because the excess flooding produced a "grassy" crop with small stalks. Myrick believed that the appearance of the rice, instead of a set flood schedule, dictated the amount of time the crop should spend in the water, "for much depends on the weather, and the order the lands are in," he observed. "As long as the rice thrives and looks perfectly green, the water can be kept on [the fields]," Myrick advised fellow planters, "but as soon as I find that it begins to get a little yellow, and not improving, I let off the water as soon as possible, always observing to change the water previous to letting it off."[45]

Like Rose, I'on family members were active contributors to the experimentation of rice cultivation. The evolution of Clayfield rice fields portray how three generations changed water management strategies from simple reservoirs and fields to elaborate irrigation networks. Prior to the Revolution, Richard I'on's limited acreage only tapped into a portion of Wappetaw Swamp. The wetlands flowed in a crescent shape around the

[44] House, *Planter Management*, 33–34, quote: 33; [Washington and Rose] "Queries on the Culture of Rice," 166; [Rose] "On Open Rice Planting," 370; Columella, "Answers to Queries on the Culture of Rice," 225.

[45] [Washington and Rose] "Queries on the Culture of Rice," 166; [Rose] "On Open Rice Planting," 370; Columella, "Answers to Queries on the Culture of Rice," 225; Jordan Myrick, "On Rice," *Letters and Extracts on Agriculture, Published by the Order of the Agricultural Society of South Carolina* (Charleston: A.E. Miller, 1824), 314–315. Point flow described in S.A. Knapp, "Rice Culture," *Farmers Bulletin* 417 (Washington, D.C.: United States Department of Agriculture, 1910), 15; Stewart, "*What Nature Suffers to Groe,*" 110; Dusinberre, *Them Dark Days*, 32.

highlands, requiring I'on to use available water from his upstream neighbor to irrigate his rice fields. His upper fields followed the contour of the high ground along the northwest corner of the property. Richard dammed two creeks to form reservoirs. Field sizes and shapes, however, were inconsistent with small squares adjoining disproportioned rectangles. In their survey of rice plantations along the East Branch of the Cooper River, anthropologists Leland Ferguson and David Babson explain that the smaller field sizes represent the early stages of either inland or tidal culture. The small size gave the planter "greater flexibility in flooding the appropriate amounts of water," according to Ferguson and Babson, where pioneering planters would have still been attempting to figure out the amount of water control for each field.[46] As one followed the field shapes further down stream and away from the settlement, the fields became more consistent in shape and size, forming elongated rectangles in the same direction as the natural water flow. By the edge of the southern boundary of Richard's Swamp Plantation, irrigation canals were parallel with cross ditches and dividing embankments segregating the field network into manageable proportions. This field shape explains how I'on felt comfortable flooding larger acreage with a higher volume of water. A combination of impounded reservoirs holding more water with growing irrigation canals gave I'on the ability to direct water to individual fields, when needed, to quickly flood the crop at required times.[47]

By observing for almost two decades his father's slaves cultivate rice, Jacob B. I'on, Sr. expanded the fields further into Wappetaw Swamp. When I'on purchased the Bruce tract in 1765, he doubled the inland rice field acreage. The southern tract allowed I'on to cultivate rice downstream further and coordinate efficient water flow through a series of straight canals, ditches, and embankments. The Bruce tract's rice fields were more elongated and uniform compared to his father's fields. Part of the field uniformity relates to the sprawling Wappetaw Swamp as it wraps around the western half of the former Bruce tract. I'on also added an additional reservoir to flood the lower fields. The water percolated from the aquifers in the upper crust of the Princess Anne Terrace. These springs percolated through the stratigraphy less than forty feet deep to form small streams merging into the Wappetaw Swamp basin. The Pleistocene

[46] Leland Ferguson and David Babson, *Survey of Plantation Sites along the East Branch of the Cooper River: A Model for Predicting Archaeological Site Location* (Columbia: University of South Carolina, 1986), 24.

[47] Wood, *Site Evaluation on Three Sites*, 15, 20.

deposits of sand, shell, and clay created a permeable stratigraphy that retained water like a sponge over the millennia, similar to the porous limestone retaining water in the Floridan aquifer, and discharged the water into streams throughout the Lower Coastal Plain. By damming these streams with embankments, I'on took advantage of the groundwater flow and retained reservoirs from a limited stream length. I'on was able to build three reservoirs within a confined landscape, and like Fairlawn, each flooded a specific field section.[48]

For plantations with little elevation change, dry highland areas were as valuable as the low-lying rice fields. The 140 enslaved people who cultivated and maintained Clayfield lived and worked within the natural boundaries of the Pleistocene swell ridges and the lowland deposits. Slave and big house settlements were ten to fifteen feet above sea level and less than sixty feet from the rice fields. The slave settlement consisted of ten houses divided into two parallel rows facing each other, which sat on a ridgeline bisecting the plantation avenue from the rice fields. The enslaved families lived approximately 200 feet southeast of the planter settlement. A low-lying ditch separated the two living areas creating a physical and, possibly, a psychological barrier between the free and the enslaved.[49] Clayfield's enslaved community settlement, like many documented by anthropologist John Michael Vlach, sat alongside the plantation's central avenue with the intention of displaying I'on's wealth to the passersby. While planters used slave housing as a representation of power, the settlements' separation from the big house enabled the enslaved to create their own culture separate from their enslavers. "Creation of a slave's landscape was a reactive expression," noted Vlach, "a response to the plans enacted by white landowners."[50] Like the planter's family, the slave community also lived within close proximity to the rice fields. With two rows of houses paralleling the fields, the closest row sat only forty feet away from the outlying embankments.

[48] "Plan of a Plantation called Clayfield," 1816; Walter R. Aucott and Gary K. Speiran, "Ground-Water Flow in the Coastal Plain Aquifers of South Carolina," *Ground Water* 23 (November/December 1985): 738.

[49] Wood, *Site Evaluation on Three Sites*, 19, 46; J.W. Joseph, "Building to Grow: Agrarian Adaptations to South Carolina's Historical Landscapes," in *Carolina's Historical Landscapes: Archaeological Perspectives*, eds. Linda F. Stine et al. (Knoxville: University of Tennessee Press, 1997), 46–48.

[50] John Michael Vlach, *Back of the Big House: The Architecture of Plantation Slavery* (Chapel Hill: University of North Carolina Press, 1993), 1, 21.

Nestled into the base of the Mt. Pleasant Scarp, Clayfield's reservoirs lay on the eastern boundaries of the rice fields. The east–west orientation of these reservoirs accented the Pleistocene ridges and troughs that formed the topographical swells. The distinctive bay galls formed finger-like wetlands extending into Wappetaw Swamp. Named after a predominant tree in the area, bay galls were the long and narrow swamps extending in a linear formation for several miles lying in between the highland swells. These topographical features provided the foundation for Clayfield rice cultivators to grow their cash crop successfully because the galls provided natural funnels for water to flow from springs and creeks down to the Wappetaw Swamp. Slaves simply had to embank the gall to adequately impound water. On the opposite side of the reservoir dams, I'on's enslaved labor force maintained the expansive field system sprawling into Wappetaw Swamp. They managed water in the individual field divisions through canals stretching the length of the galls and released the water into the downstream Fairlawn and Capers tracts. Depending on field location, water released from Clayfield rice fields went into three separate locations. Impounded water either went from the upper rice fields to Fairlawn's uncleared "swamp land," from the lower fields into the Capers Plantation reservoir or directly into the Capers Plantation eastern fields.[51]

Like Fairlawn, Clayfield's expansion toward the wetlands shows how soil and topography contributed to the development of cultivation strategies. By 1824, Jacob B. I'on, Jr. consolidated 2,814 acres that encompassed the eastern half of the Wappetaw headwaters. Using three reserves totaling 289 acres, I'on balanced irrigation with rice output on his 500 acres of embanked land. Clayfield's name came from the "stiff blue clay" found in the fields, which resembled the soil described at Fairlawn and Charleywood. The clay loam at Clayfield required slaves to "pulverize the soil" so that the composition could break down soft

[51] Wood, *Site Evaluation on Three Sites*, 19; "Plan of a Plantation called Clayfield," 1816; "Plan of a Plantation or Tract of Land in the Parish of Christ Church," 1807; "Plan of Fairlawn Plantation," 1794; "Return of Jacob B. Ion, Taxable Property in the Parish of Christ Church," 1824, SCDAH; "Return of Jacob B. Ion, Taxable Property in the Parish of St. Thomas and St. Dennis," 1824, SCDAH; [Legaré] "Account of an Agricultural Excursion," 354. The term gall refers to the gall berry bushes that grow in the swamp. I'on swamp has a noted ornithological history, as Rev. John Bachman discovered the warbler that is his namesake in the Clayfield gall in 1833. In the twentieth century, the Bachman Warbler was again sited in I'on Swamp after speculation of its extinction, see: Brooke Meanly, *Swamps, River Bottoms, and Canebrakes* (Barre, MA: Barre Publishers, 1972), 67. Estherville Plantation in Winyah Bay, Georgetown Co., SC, is another example of an inland plantation using bay galls for reservoirs.

enough for cultivation. I'on slaves – like the Fairlawn enslaved – used plows and harrows to break up the stiff clay in the central field sections, but had to break down soil with hoes near banks and drains so that the field work would not erode the features. I'on also ordered his enslaved field hands to plow old rice stubble into the soil before planting. Citing Sir Humphrey Davy's *Elements of Agricultural Chemistry*, I'on observed that the basic principles of nourishing soil depleted from farming came from introducing "manure" (compost) into the rotation.[52]

Jacob B. I'on, Jr., like Hugh Rose, paid attention to how his crop responded to the particular characteristics of soil and water. I'on maintained a similar irrigation schedule compared to Rose, with both planters eliminating the point flow from their cultivation schedule, so the planters used impounded water as efficiently as possible. I'on did not share Rose's experimental philosophy, opting to practice the traditional method of planting seed and elongating his cultivation season. The difference in sowing strategies did not dramatically effect the crop's output, as Legaré noted that I'on's "grain produced is of the first quality." Legaré continued by stating, "there are several plantations on this swamp, and [Clayfield] is one of the few where the culture of rice has been continued and found profitable."[53] Sowing seeds two to three weeks later than Rose, I'on "planted on a string," meaning that slaves laid string along the raised beds for uniformity, dropped the seeds in holes that they drilled into the beds, and covered the holes with soil before flooding the fields. While Rose saw advantages to open seed planting, I'on believed that his clay loam was not conducive to the agricultural method. I'on insisted that "leaving [rice] uncovered and flowing for the purpose of covering, and at the same sprouting, will answer on clay lands: certainly not [mine]." While both Fairlawn and Clayfield rice fields were the same soil composition, a mixture of Santee clay loam and Meggett clay loam, the variance in cultivation strategies explains how each planter put great pains into preparing the land.[54]

Like field preparation, inland planters' cultivation schedules revealed the delicate balance between water and soil management. Planter

[52] "Plan of a Plantation called Clayfield," 1816; "Return of Jacob B. Ion, Christ Church," 1824; "Return of Jacob B. Ion, St. Thomas and St. Dennis," 1824; [Legaré] "Account of an Agricultural Excursion," 354.

[53] [Legaré] "Account of an Agricultural Excursion," 354.

[54] "Plan of a Plantation Called Clayfield," 1816; "Return of Jacob B. Ion, Christ Church," 1824; "Return of Jacob B. Ion, St. Thomas and St. Dennis," 1824; [Legaré] "Account of an Agricultural Excursion," 354.

decisions were based on how much water they had at their disposal and the soil content of the individual field divisions. Because of limited reserve water, both I'on and Rose eliminated point flows to decrease one flood stage between the sprout flow and the harvest flow. Also, both planters remained observant of flood stages, as both acknowledged that too much water would damage the crop. On the final flowing, the harvest flow, I'on noted that longer flooding could damage the crop's stalk strength. He also believed swiftly flowing water could damage the rice crop during early stages of growth. When releasing water after the "long flow," his trunk minders would first start with "a slow leak" to prevent erosion "until it becomes shallow, and then as fast as possible to prevent it from scalding the rice." I'on used water to eliminate competing weeds based on the conditions of the rice crop. If the crop showed distress from too much water, I'on choose to use the hoe instead of flooding to kill weeds. However, if the weather was extremely hot or dry, then I'on would use the additional flooding to remove weeds and nourish the withered crop.[55]

Francis D. Quash used a combination of water control and cultivation techniques practiced by Rose and I'on. Like Charleywood and Fairlawn, the 3,200-acre tract of Wythewood and Cypress Hedge relied on water flowing from the Awendaw Scarp. On his 1832 examination of inland rice plantations, John D. Legaré noted Wythewood had an "extensive reservoir" that was "amply sufficient" for flooding fields. The soil content was consistent with Charleywood, Fairlawn, and Clayfield, with a combination of Santee clay loam and Meggett clay loam. This foundation served as a useful soil in retaining impounded water in reservoirs or flooded fields. To alleviate the drainage problem, "and place the management as much as possible within his control" according to Legaré, Quash ordered his enslaved laborers to construct a four-mile-long canal, the central portion of the eight-mile Wythewood Canal that connected the western branch of Awendaw Creek to the Wando River headwaters. The Wythewood Canal became the central drain of the Wappetaw Swamp plantations, tapering from six feet wide at Wythewood to twenty-five feet wide at Fairlawn.[56]

[55] [Legaré] "Account of an Agricultural Excursion," 354; Amory Austin, *Rice: Its Cultivation, Production, and Distribution in the United States and Foreign Countries*, US Department of Agriculture, Division of Statistics Report, no. 6 (Washington, D.C.: GPO, 1893), 20–22.

[56] Willoughby and Doar, "Solution to the 'Two-Talbot' Problem"; McCartan et al., *Geologic Map of the Area between Charleston and Orangeburg*; [Legaré] "Account of an Agricultural Excursion," 354; "Plan of Robert Quash Plantation," April 1789, no. 6197,

Wythewood hydrology varied from other inland plantations because of its unique location. Quash's rice fields were located on a "divide" sixteen feet in elevation between the Wando River and Awendaw Creek. Water released from Wythewood fields could travel toward either watercourse, depending on which side trunk-minders flowed the fields. Two parallel channels – serving as flanking canals – ran the length of the Wythewood and Cypress Hedge rice fields. By laying out two miles of flanking canals around the rice fields, Quash connected the two natural watercourses. Quash impounded creeks flowing down from the Awendaw Scarp, which was "comparably high" in relation to the rice fields, to form a series of small reservoirs. Wythewood did not have multiple reservoirs compared to Charleywood or Fairlawn, as the section of scarp within the plantation boundaries did not have the topographical undulations conducive for sprawling bays. Quash, however, made up for this deficiency by utilizing the numerous small streams flowing into the canal system. What Quash lacked in volume, he made up for in diversity. Unlike Charleywood and Fairlawn, which each used a singular canal to draw water into the reservoir, Wythewood depended upon a series of two- or three-foot-wide ditches and drains transferring water from the scarp to the elongated field network.[57]

Quash combined the cultivation strategies of Rose and I'on to meet the challenges of growing rice between the Wando and Awendaw headwaters. Quash slaves first used a plow, "working it as deep as possible," to cut through the Santee clay loam and then followed up with a harrow to make the soil more manageable. Quash then forced his field hands to dig rows with the hoe, instead of the plow, for it "makes neater work" and "trenches more regular and shallow." Shallow trenches allowed the crop to "tiller more" when covered with the heavy soil. While Rose

John McCrady Plat Collection, CCRMC; Williams et al., *Archaeological Survey of 3,438 Acres in the Coastal Area*, 214; "Elizabeth Bonneau to Francis D. Quash, Conveyance," 8 March 1819, Deed Book B9: 254, CCRMC; "Trenholm to Quash," Deed Book B13: 89.

[57] G.R. Lukesh, "Preliminary Examination of Waterway from Charleston, SC, to the North Santee River," in US Congress, House of Representatives, 66th Cong., 2nd sess., H. Doc. 524 (Washington, D.C.: GPO, 1920), 5; Frederic V. Abbot, "Preliminary Examination of Owendaw and Wando Rivers and Other Waters and Water Routes Connecting Bull's Bay and the Harbor of Charleston, South Carolina," in US Congress, House of Representatives, 51st Cong., 1st sess., H. Doc., Ex. Document 82 (Washington, D.C.: GPO, 1890), 2.

emphasized sowing seeds and I'on focused on flowing fields, Quash believed "the preparation of the soil for the reception of seed ... to be a very important operation, and considerable attention is paid to it" for successful cultivation along the Wando floodplain. Quash practiced covered planting, citing I'on's opinion that the clay was too stiff during open planting to cover the seeds with a thin layer of soil during the sprout flow. Quash remained consistent with his peers, however, in only using three flows, and eliminating the point flow, to irrigate his Wythewood crop.[58]

Despite "a strong prejudice against inland-swamp-rice plantations," according to Legaré, Lowcountry residents attributed the "heavy soil" of inland rice fields with producing a "heavier grain." Legaré considered Quash's operation superior to many of the surrounding plantations. "The quality of rice made on this place," wrote the *Southern Agriculturalist* editor, "is considered as of the very best and commands the highest price." Quash's perceived success came from the "intrinsic merits of the soil, and somewhat to the judicious management of the proprietor." Santee clay loam became synonymous with "heavy" soil, providing nutrients that enabled a heavy panicle. While the clay loam did not provide additional nutrients to the grain compared to other wetland soils, the loam did retain water consistently over the crop. Santee clay loam prevented water from seeping into the lower stratigraphy and provided an effective method for weed control, which allowed more nutrients to reach rice plants.[59] "Q.E.D." stressed in the *Southern Agriculturalist* that the "stiff" inland clay produced "the best and most beautiful grain." Capitalizing on select inland fields producing this desired grain, planters argued that inland seed produced better plants on tidal plantations. Rose believed that "to judge a seed by its weight" is to produce the best crop. John Bryan of Campvere Plantation also sought out inland seed with "a heavy pearly grain" to plant on his Cooper River fields, and "Columella" advocated that "inland seed is generally the best" being "larger and more pearly" to plant in rice fields. After viewing Wythewood, Legaré

[58] [Legaré] "Account of an Agricultural Excursion," 354.

[59] Ibid.; William A. Courtenay, *The Centennial of Incorporation* (Charleston: News and Courier Press, 1884), 80; Augustin L. Taveau, "Rice Culture," in *Report to the Commissioner of Agriculture for the Year 1867*, US Department of Agriculture (Washington, D.C.: GPO, 1868), 175; Austin, *Rice*, 19–22; Anna McClung, Dale Bumpers National Rice Center, United States Department of Agriculture, Stuttgart, AR, e-mail message to author, August 7, 2012.

concluded: "we doubt whether many river plantations of the same extent and number of workers, produce a greater net profit."[60]

ABSENTEE PLANTERS, MERGERS, AND ACQUISITIONS

I'on's sale of Clayfield to Stephen G. DeVeaux in 1840 signified a new phase of planter management within the Wando watershed. This marked the first time since the Revolution that a planter from outside the immediate area purchased one of the inland tracts. Fairlawn, Clayfield, and Wythewood turned over more frequently in the twenty years leading up to the Civil War compared to the previous sixty years. This turnover resulted from speculators attempting to turn a profit – often with diminishing results in the rice fields. Despite the wealth held by large planters through the antebellum Lowcountry, a declining rate of return on investment from rice production took place during the nineteenth century. Peter Coclanis calculates that while rice planters in South Carolina and Georgia barely turned a profit (if at all) in the 1840s and 1850s, they lost a staggering 28.3 percent in 1859 for the amount of money invested relative to their profits. He cites diminishing efficiency in labor, declining soil productivity, and dwindling availability of suitable rice lands as the key domestic factors leading to this declining return on investment. However, "the power of rice" had a substantial foothold in this region, which helps explain the continuing desire for rice cultivation even with the high degree of economic risk for planters.[61]

The power of rice cultivation drew planters to the practice, tidal or inland, even though the economic outcome was quite dismal. Antebellum Lowcountry planters placed most of their capital into crop production, backed by merchants and investors, so that the removal of that economic system would have devastating effects throughout the region. Also, with

[60] Q.E.D., "Answer to Queries of an Observer on the Culture of Rice," *Southern Agriculturalist and Register of Rural Affairs* 5 (December 1832): 629; J. Bryan, "On the Culture of Rice," *Southern Agriculturalist and Register of Rural Affairs* 5 (October 1832): 528; Columella, "Answers to Queries on the Culture of Rice," 225; [Washington and Rose] "Queries on the Culture of Rice," 166.

[61] Coclanis, *Shadow of a Dream*, 111–158, specifically 137–143, table 4-24: "Est. Annual Net Rates of Return on Investment SC and GA Rice Plantations and Farms, 1710–1860," 141, quote: 141; Peter A. Coclanis, "Distant Thunder: The Creation of a World Market in Rice and the Transformations It Wrought," *American Historical Review* 98 (October 1993): 1056; Peter A. Coclanis, "Introduction," in *Twilight on the South Carolina Rice Fields: Letters from the Heyward Family, 1862–1871*, eds. Margaret Belser Hollis and Allen H. Stokes, xxvi–xxvii; Dusinberre, *Them Dark Days*, 391–396.

elite rice planting dynasties consolidating land, the Middletons, Heywards, Manigaults, and Alstons to name a few, entrepreneurs looked at creative regions to tap into the rice market.[62]

DeVeaux, a successful planter from the Santee River, bought Clayfield as a way to diversify his plantation holdings. Clayfield provided a suitable candidate for inland development, with ample water sources, proper drainage, and intact infrastructure. DeVeaux, whose stepfather was Robert Marion (the older brother of Revolutionary War hero General Francis Marion), first inherited the upper Santee River tidal rice plantation Belle Isle. Entrenched in Berkeley County planter culture, DeVeaux married Anne Peyre of Spring Grove. They maintained two plantations while purchasing Woodlawn Plantation in 1810. Woodlawn was located in the center of the Santee long-staple cotton region of the Middle St. John's Parish in Berkeley County. With the boom of the long-staple cotton market, the DeVeauxs moved to Woodlawn where they built "perhaps the most imposing of all the historic houses in St. John's Parish."[63]

While DeVeaux grew tidal rice at Bell Island and long-staple cotton at Woodlawn, Clayfield's inland fields posed a new method of cultivation for the St. John's planter. DeVeaux's merger of Clayfield, Hampton, and Baldwin's Old Field expanded the Clayfield property to a total of 4,383 acres powered by 105 enslaved laborers. By combining the properties, DeVeaux controlled the central portion of Wappetaw Swamp. Like Clayfield, Hampton was the result of planters consolidating smaller tracts. The Wigfall family of Willow Hall assembled three tracts during the eighteenth century and Thomas Barksdale added an additional tract by 1806. When DeVeaux purchased the plantation, the 1,455-acre tract consisted of 380 acres of rice fields fed by a 120-acre reservoir that had

[62] Coclanis, *Shadow of a Dream*, 137–143. For further understanding on the social and cultural attachment to rice agriculture in the South Carolina Lowcountry, see: Joyner, *Down by the Riverside*; Dusinberre, *Them Dark Days*; Philip D. Morgan, *Slave Counterpoint: Black Culture in the Eighteenth-Century Chesapeake and Lowcountry* (Chapel Hill: University of North Carolina Press, 1998); Edward Ball, *Slaves in the Family* (New York, Farrar Straus & Giroux, 1998); James H. Tuten, *Lowcountry Time and Tide: The Fall of the South Carolina Rice Kingdom* (Columbia: University of South Carolina Press, 2010); Carney, *Black Rice*; Stewart, *"What Nature Suffers to Groe"*; Easterby, *South Carolina Rice Plantation*; Elizabeth W. Allston Pringle, *Chronicles of Chicora Wood* (New York: Charles Scribner's Sons, 1922); Childs, *Rice Planter and Sportsman*.

[63] Wood, *Site Evaluation on Three Sites*, 19; Francis M. Kirk, "Woodlawn," *News & Courier* (Charleston), 8 January 1935; J. Russell Cross, *Historic Ramblin's through Berkeley County* (Columbia: R.L. Bryan Co., 1985).

"never failed in the driest of seasons," according to Thomas Wigfall, "and affords water sufficient for three hundred acres." Like Clayfield's settlement, the Hampton settlement was on a peninsula of high ground that "command[ed] a beautiful view of the rice fields." An additional slave settlement also existed on another peninsula to the north. As an investment property, Hampton became just another piece of a plantation puzzle. After 1790, the big house settlement lay abandoned, leaving the enslaved labor force to live as an isolated community on the highland peninsula. Slaves lived at this settlement up to the Civil War and their descendants continued to live on the property up to the turn of the twentieth century.[64]

Adding to the reservoir system that already flowed into Clayfield, the DeVeaux tract worked as one unified unit independent of neighboring water control or field management. By 1850, DeVeaux produced the second highest rice crop in Christ Church Parish. His Clayfield output totaled 340,000 pounds. Only James Rose's Fairlawn output of 360,000 pounds superseded the upstream neighbor.[65]

Planters in the upper Wando area constantly tweaked their property holdings by aspiring to maximize their rice output. In 1845, Quash further extended his Wythewood estate westward by purchasing Rice Hope from John English. Paying only $600 for the plantation, Quash added an additional 150 acres of rice land to this existing property. The Wythewood Canal ran northwest of the Rice Hope fields, which emptied into the man-made waterway. Quash's consolidation of Wythewood, Cypress Hedge, and Rice Hope created a 3,800-acre plantation enterprise that produced 320,000 pounds of rice in 1850. Quash was the fourth-largest rice producer in St. Thomas and St. Denis Parish that year.[66]

64 "Plan of a Plantation called Clayfield," 1816; "A Plan of Thomas Barkesdale's Plantation on Wappetaw," 1792; "A Plan of Hampton Plantation," 1806, Plat Book A:40, CCRMC; Wood, *Site Evaluation on Three Sites*, 16, 19; "For Sale by Private Contract, A Valuable Settled Plantation, In Christ Church Parish" [Charleston], *City Gazette*, 25 January 1788: 4; "For Sale" [Charleston], *City Gazette*, 30 December 1806: 1.

65 1850 South Carolina Agricultural Census, Charleston District – Christ Church Parish, 337.

66 "Francis D. Quash to G.A. Trenholm, Release," 15 February 1853, Deed Book Y12: 475, CCRMC; "John English to Francis D. Quash, Release," 17 April 1845, Deed Book B12: 305, CCRMC; "Return of Francis D. Quash, Taxable Property in the Parish Christ Church," 1824, SCDAH; "Return of Francis D. Quash, Taxable Property in the Parish of St. Thomas and St. Dennis," 1824, SCDAH; 1850 South Carolina Agricultural Census, Charleston District – St. Thomas Parish, 339.

After Quash's death in 1853, his executors sold Wythewood plus seventy-two slaves to Charleston cotton broker and financier George A. Trenholm to satisfy Quash's $25,000 debt. Trenholm consolidated Wythewood and Willow Hall to form a 5,761-acre tract. Trenholm, like DeVeaux, was an absentee planter who invested in plantations throughout the Lowcountry. Evidently he continued Wythewood's productivity as he was singled out as one of the active inland planters by 1860. However, by this time, Trenholm had sold the entire lot to William Lucas, another successful south Santee River tidal planter who produced 1,575,000 pounds of rice from his combined properties in 1860.[67]

By the final decade of the antebellum era, the four consolidated plantations – Charleywood, Fairlawn, Clayfield, and Wythewood – all experienced significant declines in inland rice productivity. Charleywood rice fields yielded only 4,000 pounds of rice in 1860, compared to 120,000 pounds ten years earlier. St. Helena planter Benjamin J. Johnson, who purchased Fairlawn in 1857, produced 24,000 pounds in 1860. His output was 336,000 pounds lower than the amount James Rose had produced a decade earlier. Upstream at Clayfield, Thomas Wagner grew 72,000 pounds of rice, which was 20 percent of DeVeaux's 1850 crop. During this period of planting, a movement of property speculation took place that did not allow property owners much time to understand their inland environments. Of these three property owners, Wagner was the only businessman who had owned his property for longer than six years between 1850 and 1860. Lucas and Johnson, in comparison, purchased their tracts only two years before the 1860 census.[68]

The decline of these inland plantations is consistent with the general decline of output for the South Carolina rice industry between 1849 and 1859. As mentioned earlier, domestic reasons for this downward trend lay with labor productivity, scarcity of capital, and declining soil fertility.

[67] "Quash to Trenholm," February 15, 1853, Deed Book Y12: 475; Williams et al., *Archaeological Survey of 3,438 Acres in the Coastal Area*, 61; Thomas R. Wheaton et al., *Archaeological Site Testing of Willow Hall and Walnut Grove Plantations, Francis Marion National Forest*, Francis Marion and Sumter National Forests Cultural Resource Management Report 92-13 (Stone Mountain, GA: New South Associates, April 22, 1992), 42; 1860 South Carolina Agricultural Census, Charleston District – St. James Santee, 329; Charleston Mayor and historian Courtenay stated in 1883 that, "some inland rice fields were in use as late as 1860, as for instance Mr. Trenholm's 'Wythewood' plantation in St. Thomas Parish"; Courtenay, *The Centennial of Incorporation*, 80.

[68] 1850 South Carolina Agricultural Census, Charleston District – Christ Church Parish, 337; 1860 South Carolina Agricultural Census, Charleston District – Christ Church Parish, 315.

During this ten-year period, as William Dusinberre calculates, the South Carolina enslaved rice laborers declined by as much as one-sixth. Planters transferred their enslaved labor to grow sea-island cotton or work on more fertile tidal rice plantations developing in Georgia and North Carolina. For South Carolina rice planters, a decline in rice output lay with the soil. After decades of intensive cultivation, even with the help of alluvium fertilizing the fields, both tidal and inland plantations finally reached a tipping point in output. For inland planters, their inland rice fields could not keep up with the hard practices of annual cultivation without field rotations or rest.[69]

Combining increasing water control projects and an ever-expanding enslaved labor population with an established Lowcountry rice market economy and emerging tidal irrigation technology, inland rice field practices had changed dramatically by the eve of the Civil War. Studying places like Charleywood, Fairlawn, Clayfield, and Wythewood reveals the ecological intricacy of these plantation systems and the larger history of inland rice cultivation. While large-scale inland rice cultivation varied from tidal cultivation in terms of how people irrigated the fields, the two cultivation strategies were not separate. Rice planters in general suffered from problems of diminishing returns and the ability to manage labor and the natural landscape. "Rice planting was … an immensely complex business," observes Dusinberre, "fraught with unexpected crises, it required long experience and depended on the managers' acquiring numerous and varied skills, and on their exercising fine judgment."[70]

This story of large-scale inland rice cultivation reveals that the plantations were not static. In their most mature form, inland rice plantations resembled their tidal counterparts in terms of extensive irrigation networks, labor control, and agricultural schedules. What separated the two systems was how they handled the natural resource of water. Inland planters had to ask different questions, dealing with impounding water and using the limited resource to flood fields. Both forms of rice cultivation not only required that cultivators maintained a critical understanding of how to grow rice, but also how to utilize the surrounding landscape to the best of their ability. Inland planters on the upper Wando River used a

[69] Coclanis, *Shadow of a Dream*, 137, 283–284, n. 65; Dusinberre, *Them Dark Days*, 388–389.

[70] Dusinberre, *Them Dark Days*, 12.

variety of microenvironments to harness water and cultivate fields. Brackish tidal floodplains, bay galls, and small-stream floodplains were zones that these inland planters manipulated. Tidal planters, on the other hand, only used the "tidal zone" to grow rice. Like their tidal counterparts, inland planters had to control water through floodplains, yet not fall victim to natural disasters such as freshets or droughts. This story moves beyond how people planted the crop, but also how they shaped the land within the constraints of specific microenvironments.

5

"The Rice Fields which Are Sown Have Been Partially Flowed"

Water and Labor Management during the Antebellum Period

On New Year's Day in 1845, Mathurin Guerin Gibbs looked forward with optimism to cultivating rice on his newly acquired inland fields. The slaves "are preparing the rice lands for me," Gibbs wrote, "and will afford me an advantage which [I] have never before enjoyed."[1] The planter at Jericho Plantation, located on the headwaters of the East Branch of the Cooper River thirty miles northeast of Charleston, anticipated a successful harvest after moving his family across the Cooper River the previous year.[2] Unlike tidal fields with their abundant and consistent water supply, Jericho's inland fields were at the mercy of reservoirs. Concern for water flow and availability appear throughout Gibbs' ledger book for the next four years. In 1845, he began the agricultural cycle by ordering two slaves to clear the drainage canal for efficient water flow from the reservoir dam to the adjacent rice fields. In addition, enslaved men prepared the rice field by "chopping," or aerating, both soil and leftover stubble before sowing seeds, a task that would continue for the next two months. While observing this laborious activity, Gibbs noted that slaves' work would "clear the water course so to admit an easy flow of water into the fields sown."[3] Two days later, "one hand was attending the flow of water from Hell Hole [Swamp]."[4] To irrigate the fields effectively, planters had slaves level any high ground that would cause dry spots when flooded; also

[1] 1 January 1845, Mathurin Guerin Gibbs Plantation Register, South Carolina Historical Society (SCHS), Charleston, SC.

[2] Ibid., 8 May 1843. [3] Ibid., 16 April 1845. [4] Ibid., 18 April 1845.

they filled in any indentations in the ground, which, if left, would create bogs after the fields drained. After his slaves sowed the fields, Gibbs realized the reservoir bank was leaking water in his smaller field, called "Pipkin." This leak put more water in the Pipkin field than Gibbs anticipated, and as a result, depleted the reservoir water needed for his second field, "Jim New Grounds." After debating whether he had enough impounded water for the season, Gibbs gambled and "ordered water to be run off so as to let the last sown rice come up ... [as the reservoir] has sufficient strength to flow Jim New Grounds."[5] Balancing the management of water between the two fields would become a constant issue for the rest of the season.

One month later, Gibbs remained worried about his reservoir situation. An ensuing drought limited Hell Hole Swamp's water flow to the Jericho reservoir. In order to stay within the cultivation schedule, the trunk minder began the long flow on Pipkin "with all of the water that could be got from the fallow fields and the big-dam reserve from Hell-Hole."[6] Without the proper water depth, Gibbs' rice would not grow to a sufficient height. Two days after the flooding, he wrote: "the rice fields which are sown have been partially flowed. It will require more rain to flow sufficiently."[7] Gibbs' trunk minder had to channel water from nearby fields through an intricate canal system interwoven through his plantation. These canals provided the arteries through which rice planters could pump water to specific areas.

Because inland planters could not rely on consistent water sources to flood fields like their tidal counterparts, planters and slaves became innovative water managers. At the same time, they acknowledged their environmental limitations. Inland rice planters could only irrigate their fields with available water from impounded reservoirs. In times of heavy rain, this impounded water proved too much and breached dams and embankments. In times of drought, reservoirs dried up, leaving planters with little water to irrigate their rice crop and eradicate weed-like grasses. To make water work for them instead of against them, inland rice planters devised canal and drain networks that fit within the plantation's rice field topography. With managing water flow on and off the rice fields, inland planters understood that this natural resource was limited. Inland planters could not afford the luxury of flooding fields four times a season, as their tidal peers did. They also realized that they could not produce the

[5] Ibid., 28 May 1845. [6] Ibid., 26 June 1845. [7] Ibid., 30 June 1845.

same volume of output compared to the larger, more efficient tidal systems. On average, inland planters generated ten bushels an acre less than their tidal contemporaries. For Gibbs, the labors of growing inland rice would often lead to "unprosporous" results, but each planting season would bring new optimism.[8]

Gibbs' story represents the limitations of resources found in antebellum inland planting and how planters aspired to overcome these shortcomings. Limitations came in a variety of examples: the lack of available water adequate crop irrigation, the lack of labor needed to manage such an intensive operation, and the lack of social privilege to inherit more desirable lands. Despite these limitations, planters looked to African-American expertise to successfully grow rice. Just as this reliance took place one hundred years prior, planters with limited resources to hire white overseers utilized the guidance of the enslaved to help manage the temperamental water control strategies and critical cultivation schedules. As discussed later in the chapter, Gibbs did not have the background to personally dictate to laborers how to grow rice, nor did he have the economic means to pay for a knowledgeable overseer. With limited labor and resources, growing rice reveals the temperamentally of these microenvironments.

Gibbs and his neighbour John Beaufain Irving were two professionals, one a lawyer and the other a doctor, who saw inland rice property ownership and agricultural output as a way to elevate their social standing at a time when tidal cultivation was beyond their reach. In "status-conscious" Charleston, planters represented the aristocratic class.[9] According to southern historian Jennifer Goloboy, "planters sought to consolidate their social power through their claims to refinement." Aristocracy, expressed through "family connections, landed wealth, agreeable manners, sociability, and conspicuous leisure," served to segregate the haves and the have-nots. As a group, professionals and merchants, regardless of their economic successes, fell into the bourgeoisie. For those - upper-middle class people owning established plantation tracts,

[8] Ibid., 1 June 1846.

[9] Quote: Dale Rosengarten and Theodore Rosengarten, eds., *Portion of the People: Three Hundred Years of Southern Jewish Life* (Columbia: University of South Carolina Press, 2002), 102. The Cohen family owned two inland rice plantations along the Ashley River in St. Andrew's Parish during the antebellum period. Marx E. Cohen's Clear Springs plantation had twenty-five people working a thousand acres, which produced rice, corn, livestock, timber, and bricks.

landholding signified both an arrival of economic prosperity and an attempt to breach the exclusive social circles.[10]

Gibbs and Irving represented the economic limitations seen in other merchants, artisans, and professionals striving to advance their social status. They epitomize "middle-class masters," those men who purchased land and slaves in their attempts to emulate the planter aristocracy. Usually, middle-class plantation owners were well educated, owned between five and fifty slaves, and used agriculture as a means to supplement their income and status. As historian James Oakes discusses in *The Ruling Race*, "it has always been one of the defining characteristics of middle-class slaveholders that they combined careers (of their original profession and agriculture) to enhance their prospects for upward mobility."[11] Both Gibbs and Irving reveal the complexity of Southern society. Both men were born into privilege, yet because of their birth-order, neither received the riches and leisure of planter aristocracy. Instead, both men had to seek out an occupation that served as a conduit toward the trappings of the upper-echelons. In doing so, Gibbs and Irving straddled between respectable professions, yet lowered their economic status into the developing middle class, while maintaining social privilege and status found in the planter identity.

Many antebellum inland rice planters epitomized Oakes' "middle class masters." Oakes stresses that these planters were a product of westward migration and cheap land. However, those aspiring owners of inland rice plantations in the South Carolina Lowcountry remained close to their families and social networks. In the early nineteenth century, inland rice plantations had lost worth compared to market values during the inland swamp boom before the Revolution.[12] Real estate values for non-tidal rice plantations declined through the antebellum period, from an average of \$20–\$50 per acre at the turn of the nineteenth century to \$3–\$5 per acre in the 1840s.[13] The drop in inland plantation real estate values

[10] Jennifer L. Goloboy, "Strangers in the South: Charleston's Merchants and Middle-Class Values in the Early Republic," in *The Southern Middle Class in the Long Nineteenth Century*, eds., Jonathan Daniel Wells and Jennifer R. Green (Baton Rouge: Louisiana State University Press, 2011), 52; Walter J. Fraser, Jr., *Charleston! Charleston!: The History of a Southern City* (Columbia: University of South Carolina Press, 1989), 196.

[11] James Oakes, *The Ruling Race: A History of American Slaveholders* (New York: Vantage Books, 1982), 57–65, quote: 58.

[12] Peter C. Mancall et al., "Agricultural Labor Productivity in the Lower South, 1720–1800," *Explorations in Economic History* 39 (October 2002): 410–411.

[13] Oakes, *Ruling Race*, 57–65; Joyce E. Chaplin, *An Anxious Pursuit: Agricultural Innovation and Modernity in the Lower South, 1730–1815* (Chapel Hill: University of North

stemmed from the growing demand of tidal rice plantations and also from a myriad of negative perceptions attributed to the older agricultural lands. By 1824, three key factors deterred inland rice cultivation: unreliable irrigation caused by freshets and droughts, soil deterioration from generations of cultivation, and unhealthiness of the land from "stagnant water and vegetable decompositions of abandoned rice fields." While inland rice cultivation allowed planters to attempt to improve their status without heading to the frontier, the plantations presented their own set of ecological challenges.[14]

Weakened inland property values made inland plantations more affordable to people willing to cultivate rice and strive for planter status using the reservoir method. Although Gibbs and Irving were descendants of established planter bloodlines, initially neither man could work his way into the tidal establishment because of the high capital commitment. By the antebellum period, tidal rice planters dominated the Lowcountry's economic, social, and political scene. These planters generated ever-increasing wealth, which allowed them to purchase more tidal rice lands and use their economic stature to secure in key political positions. Planters cultivated rice along established tidal corridors, and they generated tremendous output when environmental and economic conditions were in their favor. Although antebellum tidal rice planters did not see the lucrative price-per-pound ratios that their colonial counterparts did, their higher yields of rice per acre and their ability to compound capital in land and labor enabled them to maintain tremendous economic, social, and political exclusivity.[15]

Carolina Press, 1993), 247; Lewis Cecil Gray, *History of Agriculture in the Southern United States to 1860*, vol. 1 (Washington, D.C.: Carnegie Institution of Washington, 1933; reprint Gloucester, MA: Peter Smith, 1958), 405; William M. Mathew, ed., *Agriculture Geology, and Society in Antebellum South Carolina: The Private Diary of Edmund Ruffin, 1843* (Athens: University of Georgia Press, 1992), 78–79.

[14] Abraham Blanding, "Report on the Edisto Canal," [1824], Committee Reports, South Carolina Department of Archives and History (SCDAH), Columbia, SC, 376–382, quote: 379; also see: Thomas Y. Simons, M.D., "Remarks on the Climate of the Lower Country of South Carolina," *American Journal of the Medical Sciences* 9 (November 1831): 256; "Old School" described droughts evaporating reservoir water, while "the unfortunate man looks upon his reserve, now sunk to a mere puddle, and sees a picture of his fate; the frogs and fish no longer able to *dive* from the snakes and cranes, are swallowed by them alive, and while he compares these last to his creditors; he hopes they may be more merciful." "Inland Swamps," *Southern Patriot* 29 (April 1823): 2.

[15] For declining rates of return for Lowcountry rice planters during the nineteenth century, see: Peter A. Coclanis, *The Shadow of a Dream: Economic Life and Death in the South Carolina Low Country, 1670–1920* (New York: Oxford University Press, 1989),

The exclusive circle of tidal rice planters made it difficult for outsiders to obtain tidal lands. Families passed lucrative or sentimental property from generation to generation through marriage and inheritance. By the antebellum period, fortunate family members inherited desirable property that provided consistently high yields in agricultural production. Depending on the number of children subject to inheritance, a son or a daughter could inherit one or multiple properties from the family patriarch. Sons continued practicing the agricultural precedent laid out by their fathers. The inheritance of a plantation by a daughter served either as a dowry for future husbands, or as additional family capital if the daughter was already married. An intricate web of South Carolina families intermarried over the generations as offspring joined one another's families to preserve land titles. The legacy of these multigenerational landholdings included important enslaved families who understood the complexity of rice cultivation and the natural subtleties in changing conditions. Beyond owning prosperous tidal rice lands and having little, if any, debt, these planter families had the luxury of relying on experienced enslaved cultivators. For example, the Ball family maintained a hold on Comingtee, Strawberry, Hyde Park, Kensington, and Limerick plantations for over seven generations. Comingtee, the family seat, was in the Balls' possession for 229 years. Hyde Park remained in Ball ownership up to the last decade of the twentieth century, representing 253 years of family possession. During the colonial and antebellum periods, the Balls acquired twenty-three plantations along the Cooper River. Ball daughters married into the Harleston, Corbett, Laurens, Simons, Smith, Waring, Moultrie, and Rutledge families; these unions further sealed the Cooper River floodplains from outside planters. Conveyed between the generations of Balls were hundreds of slaves who practiced rice cultivation, learning the art from their elders and passing it on to their descendants.[16]

140–141. For the concentration of tidal rice plantations, see: Chaplin, *Anxious Pursuit*, 237–238; William Dusinberre, *Them Dark Days: Slavery in the Inland Rice Swamps* (New York: Oxford University Press, 1996), 191–196; Coclanis, *Shadow of a Dream*, 69–70; and James M. Clifton, "The Rice Driver: His Role in Slave Management," *South Carolina Historical Magazine* 82 (July 1981): 331.

[16] William Kauffman Scarborough, *Masters of the Big House: Elite Slaveholders of the Mid-Nineteenth-Century South* (Baton Rouge: Louisiana State Press, 2003); Jennifer R. Green, "Born into Aristocracy?: Professionals with Planter and Middle-Class Origins in Late Antebellum South Carolina," in *The Southern Middle Class in the Long Nineteenth Century*, eds. Jonathan Daniel Wells and Jennifer R. Green (Baton Rouge: Louisiana State University Press, 2011), 157–159; Edward Ball, *Slaves in the Family* (New York, Farrar Straus and Giroux, 1998), 2; for further explanation of Lowcountry rice planter

A second reason why outside, aspiring rice planters had difficulty acquiring tidal lands was the tremendous amount of capital needed to start up such a venture. Tidal cultivation represented a "class-based innovation" favoring the privileged. Established tidal plantations, each with their large population of labor to manage and maintain the enterprise, went for large sums. As Joyce Chaplin explains in *An Anxious Pursuit*, cultivated tidal swamp lands averaged $70–$90 per acre, an "indication that rice planting was no longer a possibility for men with modest resources." In 1796, Nathanial Pendleton estimated a new tidal rice plantation would cost $10,570. And forty years later, Charles Manigault invested $49,500 to start production at the Savannah River tidal plantation, Gowrie.[17] These prices exceeded greatly what people paid for non-tidal properties. For example, John B. Irving paid only $2,000 for his inland plantation in 1840.[18] Consistent with real estate, taxes on the tidal marshes were twice as high or more per acre compared to the surrounding land. Landowners paid $8 per acre in tax for tidal marsh, compared to $4 per acre of inland swamp and $1 an acre of pineland.[19]

By the mid-nineteenth century, people living across the South Carolina and Georgia Coastal Plain acquired and manipulated land, relying on the labor of thousands of slaves, to grow a crop that signified a particular prestige. Charleston's port provided opportunity for merchants to buy into the planter elite. Isaiah Moses provides a unique illustration of how a Charleston merchant realized this agrarian preeminence through purchase of an inland rice plantation. Moses, born in the Kingdom of Hanover in northeast Germany, immigrated to America before 1800 and

interconnected bloodlines, see: Anne Simons Deas, *Recollections of the Ball Family of South Carolina and the Comingtee Plantation* (Summerville, SC: Alwyn Ball, Jr., 1909); Henry A.M. Smith, "The Ashley River: Its Seats and Settlements," *South Carolina Historical and Geological Magazine* 20 (April 1919): 85, 97–98, 108–115; Henry A.M. Smith, "The Upper Ashley: And the Mutations of Families," *South Carolina Historical and Geological Magazine* 20 (July 1919): 165–168. For studies on multi-generations of Ball family slaves, see Ball, *Slaves in the Family*, and Cheryll Ann Cody, "Slave Demography and Family Formation: A Community Study of the Ball Family Plantations, 1720–1896" (Ph.D. diss.: University of Minnesota, 1982).

17 Chaplin, *Anxious Pursuit*, 247, 238–239; Dusinberre, *Them Dark Days*, 16; Mancall, "Agricultural Labor Productivity," 410–411.

18 John B. Irving, "Record of Windsor and Kensington Plantations, 1840–1888," Charleston Library Society (CLS), Charleston, SC; "John B Irving, Mortgage of Plantation," 23 May 1841, Deed Book H11:57, Charleston County Register of Mesne Conveyance (CCRMC), Charleston, SC.

19 "An act to raise supplies for the year one thousand eight hundred and four, and for other purposes therein mentioned," *City Gazette* (Charleston), 9 January 1805: 2.

established himself as a Charleston merchant within the first decade of the nineteenth century. He worked his way up the economic ladder, first as a grocer then later as owner of a dry-goods business on King Street. By 1813, he purchased The Oaks, an inland rice plantation located on Goose Creek. At the time Moses purchased the property, the sixty acres of inland fields were worn out, yielding little after a century of heavy use. Moses paid $6,000 for 794 acres, or $7.55 per acre, well under the average $20–$50 per inland swamp acre at that time and the much higher tidal acre costs, as noted by Chaplin.[20]

The Oaks received its name from a grand avenue of live oaks one-third of a mile long, "across which the branches of the trees interlock," leading to the plantation settlement. The Middleton family owned this property for 116 years, beginning in 1678 when Edward Middleton received the original land grant. Middleton's sale of The Oaks, their family seat, in 1796 represents the shifting land values after the Revolutionary War. As documented by Chaplin in *An Anxious Pursuit*, a dramatic land exchange began at the turn of the nineteenth century with the onset of people attempting to take advantage of the new tidal technology. Before the advent of soil sciences and commercial fertilizers, heavily cultivated inland fields like those at The Oaks suffered from nutrient depletion.[21]

With The Oaks purchase in 1813, Moses saw this established plantation as a mode of entry into planter aristocracy. From 1796 to 1813, the plantation changed hands twice with the speculative real estate market of the time. For Moses, the plantation's allure came from its connection to the Middletons, a family that represented the upper echelon of the planter elite. By 1819, Moses changed his title from "merchant" to "planter" in the Charleston city directory. Although Moses changed his title, he wisely supplemented his agricultural income with his dry-goods business. While the profitability of The Oaks during Moses' tenure is unclear, he did increase the plantation's enslaved population from thirty-five people in 1830 to fifty people a decade later. He and his wife, Rebecca, remained active in the city's thriving Jewish community, where Moses was a significant financial supporter and served on the governing board to the Kahal Kadosh Beth Elohim (KKBE) Congregation. Moses maintained

[20] Rosengarten and Rosengarten, *Portion of the People*, 103–105; Michael J. Heitzler, *Goose Creek: A Definitive History*, vol. I: *Planters, Politicians, and Patriots* (Charleston: The History Press, 2005), 211; Chaplin, *Anxious Pursuit*, 247, 238–239.

[21] Quote: Mathew, *Private Diary*, 62; Rosengarten and Rosengarten, *Portion of the People*, 103–105; Heitzler, *Goose Creek*, 211; Chaplin, *Anxious Pursuit*, 247.

the property for twenty-seven years, but financial difficulties forced him to sell the property in 1840 at a loss of $2,000.[22]

Gibbs and Irving, the lawyer and the doctor, were two aspiring planters who reached two different outcomes. Gibbs struggled as an inland planter for ten years before dying from complications of malaria in 1849. Irving, on the other hand, used his inland rice plantation as a springboard, and by the eve of the Civil War, he had obtained larger tidal estates. During the time that Gibbs and Irving planted inland rice, their annual output varied from a total loss to moderate profits. The subtle relationships between the spatial positioning of plantations, topographic formations affecting water flow, and the amount of labor necessary to control this water flow, reflect how precarious the inland landscape could become for these two planters and their slaves.[23]

Mathurin Guerin Gibbs provides an example of one individual attempting to gain status through inland rice cultivation. Gibbs, born into a prominent Charleston family, had the advantage of social positioning from birth. His father, Joseph Gibbs, emigrated from Bermuda and became a planter on the Stono River in St. Andrew's Parish. The family's rural presence was short lived, as Joseph died when Mathurin was three years old. Susana Guerin Gibbs abandoned the Stono estate and moved her family to Charleston. Mathurin graduated from law school and married Maria Louisa Poyas, daughter of planter John Ernest Poyas.[24]

While practicing law in Charleston, Gibbs sought to enter the planter class by purchasing Rice Hope Plantation in 1837 for $4,000. Like The Oaks, Rice Hope was an inland rice plantation located on Goose Creek in St. James Parish.[25] Although the 1,007-acre property had the infrastructure to support the crop, Gibbs did not grow rice as a commodity. Instead, he focused on Santee black seed cotton, which (as will be discussed in the next chapter) was a popular cash crop grown above former inland

[22] Rosengarten and Rosengarten, *Portion of the People*, 103–105; Judith Alexander Weil Shanks, *Old Family Things: An Affectionate Look Back*, 15, 17, www.judithwshanks.com/old_family_things/index.html (accessed April 22, 2019); Heitzler, *Goose Creek*, 211.

[23] Gibbes Family History and Genealogy Research Files, SCHS; Dr. John Beaufrain Irving Family Papers, SCHS.

[24] Ball and Gilchrist Family Papers, MS vol. 43, 1923–1934, South Caroliniana Library, University of South Carolina (USC); Gibbes Family History and Genealogy Research Files, SCHS.

[25] St. James Goose Creek is not to be confused with St. James Parish bordering the Santee River, also called St. James Santee. Both parishes were created in 1706.

rice fields in Middle St. John's Parish. From 1838 to1844, Gibbs devoted only two acres each year to rice cultivation, which produced poor yields of five to eight bushels per year in comparison to the thirty to fifty bushels per acre grown by inland rice planters during the antebellum peiod. A series of droughts and thus compromised harvests, beginning in 1838 and lasting until 1842, prevented Gibbs from paying off his mortgage. Eventually the crop failures led him to bankruptcy and forced him to sell the plantation at auction in February 1843. Despite Gibbs' failure as a planter, he paid rent and continued to live on the property until 1844, when he and his family moved to Jericho.[26]

After leaving Rice Hope, Gibbs retired from law and continued his gentleman planter lifestyle. It was at Jericho that Gibbs made a grievous error when he ended his legal profession. As Oakes emphasizes in *The Ruling Race*, having two careers "protected the slaveholding middle-class from the most severe instabilities of the agricultural economy."[27] The money from a dual career was invaluable to the aspiring planter, as it provided insurance for poor crop harvests or declining market prices. Gibbs' desire to enter the planter class motivated him to make a decision that proved shortsighted. He left the legal profession to focus all of his efforts on agriculture. Jericho was a temperamental plantation because of the difficulty of controlling water there. By devoting all of his efforts to cultivating rice, Gibbs set up his family for economic failure with no primary career to fall back upon during times of hardship.

Jericho's temperamental reputation had to do with its location at the headwaters of Nicholson Creek and its reliance on Hell Hole Swamp for irrigation.[28] The northern portion of the plantation lay above the Bethera Scarp, on a linear ridge of upland pine and oak forests, which is a former barrier island system that runs parallel to the South Carolina coastline. The elevation across the scarp drops twenty feet in elevation to a bowl of clayey loam soils with cypress and hardwoods. The geological formation of high sandy ridges, in close association with low-lying clay loam,

[26] Deed Book R10:468, Deed Book L11:128, CCRMC; 13 December 1842, Gibbs Plantation Register, SCHS; Ball and Gilchrist Papers, MS vol. 13, 19 December 1844, USC.

[27] Oakes, *Ruling Race*, 58.

[28] Donald J. Colquhoun, "Cyclic Surficial Stratigraphic Units of the Middle and Lower Coastal Plains, Central South Carolina," in *Post-Miocene Stratigraphy Central and Southern Atlantic Coastal Plain*, eds. Robert Q. Oaks and Jules R. DuBar (Logan: Utah State University Press, 1974), 189.

became central to Jericho's rice cultivation and settlement patterns.[29] Almost always, scarps form exclusive demarcations between each of their associated coastal plain terraces with older barrier islands on the landward side and younger back-barrier systems on the seaward side. In some areas the elevation differences across the scarp are small – as little as five feet. In other areas the differences can be large, such as twenty feet across the aforementioned Bethera Scarp. The barrier island system at Bethera is significantly higher than the surrounding area, up to forty-feet higher than the next younger back barrier system, with a peak elevation of sixty feet. Then it descends gradually in northeastern and southwestern directions to thirty-five and forty feet, respectfully. These changes in elevation along its length occur over approximately twelve miles.[30]

Bountiful access to water was a critical component of success for inland rice planters. Hell Hole Swamp and Jericho Swamp, both located to the south east of the Bethera Scarp, provided the crucial water source for Jericho and Windsor planters. These wetland areas were part of the Pleistocene estuarine (back barrier) flats that formed a hinterland of approximately twenty square miles. The two swamps are interconnected, low-lying landforms sandwiched between the linear highland and the Santee delta. The swamps act as broad watersheds, draining the upland topography and forming the headwaters of Nicholson and Turkey Creeks.[31] Originally, the swamps' clay loam supported cypress and tupelo gum stands, each "like a cloud hovering over the horizon in the distance" according to Gibbs. Later these wetland areas became the central location

[29] Cable, John S. et al., *An Archeological Survey of 3,720 Acres in The Bethera Area, Wambaw and Witherbee Districts, Francis Marion National Forest*, Francis Marion National Forest Indefinite Services Survey Report 2 (Columbia: United States Department of Agriculture, Forest Service, 1991) 20; David Anderson and Patricia A. Logan, *Francis Marion National Forest Cultural Resources Overview* (Columbia: United States Department of Agriculture, Forest Service, 1981), 1, 4; Donald J. Colquhoun, *Terrace Sediment Complexes in Central South Carolina* (Columbia: University of South Carolina, 1965), 23–27; Donald J. Colquhoun, *Geomorphology of the Lower Coastal Plain of South Carolina* (Columbia: Division of Geology, 1969), 23, 6.

[30] William Richardson Doar, III, and Christopher George St. Clement Kendall, "An Analysis and Comparison of Observed Pleistocene South Carolina (USA) Shoreline Elevations with Predicted Elevations Derived from Marine Oxygen Isotope Stages," *Quaternary Research* 82 (July 2014): 165–167; W.R. Doar, III, "Advice," email, 2018; Colquhoun, "Cyclic Surficial Stratigraphic Units," 181; United States Department of Agriculture, *Soil Survey of Berkeley County, South Carolina* (Washington, D.C.: GPO, 1980), 43, 44; Colquhoun, "Cyclic Surficial Stratigraphic Units," 180–182.

[31] Robert E. Weems and Earl M. Lemon, Jr., *Geology of Bethera, Cordesville, Huger, and Kittredge Quadrangles, Berkeley County, South Carolina*. Washington, D.C.: United States Geological Survey, 1989.

for inland rice cultivation, as the large quantities of water flow made this environment suitable for cultivation. At the same time, the broad bays that form Hell Hole and Jericho Swamp presented planters with a difficult assignment: to manage water flow adequately through, in some opinions, a featureless topography that could not provide substantial borders for irrigation canals and field embankments.[32] Jericho plantation encompassed a dramatic yet subtle change in geology. Within a few acres was a diverse array of swamps, floodplains, stream terraces, bays, upland flats, slopes, and ridges.[33] Jericho's role as a secondary plantation during the eighteenth century represents how planters, according to S. Max Edelson, "supported the diversification of core plantations and the specialization of frontier plantations."[34] In the case of the Balls, Jericho provided additional income for the family and specific foodstuffs for their plantations. Francis Harris owned 500 acres of the Jericho tract in 1718–1719. Eventually Nicholas Mayrant and his wife Susanne came into possession of the property, conveying it to Daniel Huger before 1754. Benjamin Huger, Daniel's son, inherited the property and sold it to John Coming Ball.[35]

By 1783, the Ball estate had up to thirty-eight slaves on the property. John Coming Ball's son, John Coming Ball II, used the property for agricultural purposes to complement his other inland rice plantation, Back River, in St. James Goose Creek. Representative of the Lowcountry plantation economy, the Jericho enslaved labor force grew rice, raised cattle, and manufactured cypress shingles. In 1783, John Coming Ball II sold 111,900 shingles to neighboring planters beginning to rebuild after the Revolutionary War.[36] The Jericho cattle population varied between 100 head in 1788, 188 cattle in 1789, and 111 cattle in 1793.[37] After John Coming Ball II's death in 1792, Jericho received little attention for almost two decades. During that time, Ball's cousin balanced care of both

32 20 January 1845, Gibbs Plantation Register, SCHS; Cable et al., *Archeological Survey*, 20.

33 Anderson and Logan, *Francis Marion National Forest Cultural Resources Overview*, 6.

34 S. Max Edelson, *Plantation Enterprise: Plantation Enterprise in Colonial South Carolina* (Cambridge, MA: Harvard University Press, 2006), 224.

35 Cable et al., *Archeological Survey*, 86.

36 John Coming Ball Account Book, "Account of Shingles, Lumber, ect Sold from Jericho Plantation for the Year 1783," Ball and Gilchrist Papers, MS vol. 15, 1783–1810, USC.

37 "Livestock, 1788–1854," John Ball, Sr., Ball Family Papers, SCHS; "Appraisement of the Stock at Jericho Plantation Belonging to the Estate of John Coming Ball, Deceased," Ball Family Papers, SCHS.

Jericho and Back River with his own plantations. The Ball family leased Jericho to John Jaudon for £15 a year between 1797 and 1803; then the plantation fell into disrepair until Isaac Ball took control in March 1810.[38]

After Isaac Ball inherited Jericho in 1810, he transformed the tract from a secondary plantation to a central plantation during his tenure. In similar fashion to the upper–Wando River planters, Isaac pieced together a massive 14,489-acre landholding that stretched eight and a half miles through the Huger Creek watershed between 1810 and 1825.[39] Jericho, annexed with Nicholson and Farewell Corner Plantations, formed a 5,805-acre inland complex.[40] By combining these three plantations, Isaac Ball gained control of the Nicholson Creek headwaters and floodplains. Isaac's assembly of this network resembled his father's creation of Midway Plantation, a 2,421-acre rice plantation cradled between the two branches of the Cooper River. By 1798, John Ball purchased five tracts of land to surround a Carolina bay, which was an oval-shaped depression of standing water that was, in this case, one mile long. Water did not flow freely through the depression, making inland rice culture difficult to manage. However, John Ball constructed an extensive canal network to irrigate the 250 acres of rice fields and to draw water off the bay. Isaac managed Midway between 1802 and 1810, gaining critical knowledge in regard to managing slaves and understanding the inland agricultural cycle.[41]

A detailed understanding of inland rice output puts in context the tedious nature of obtaining desirable yields in the context of specific microenvironments. As seen, the output in a specific location varied and successful planters had to rely upon a large labor population in order to maintain a stable agricultural infrastructure. For example, Isaac Ball's use of Jericho slaves to grow rice reflects the strong market during the early nineteenth century. Between 1812 and 1824, Jericho's slave population

[38] Ball and Gilchrist Papers, MS vol. 15, 1783–1810, USC.

[39] "Tax Return, Isaac Ball, 1824," Ball Family Papers, SCHS; "Memorials and Misc, 1801–1833," Ball Family Papers, SCHS; "Release and Plat of James Whitesides to Isaac Ball, 23 February 1814," Deed Book H8: 217, CCRMC.

[40] "Tax Return, Isaac Ball, 1824," SCHS.

[41] "Elias Ball Muniments," Ball Family Papers, SCHS; Mark S. Schantz, "'A Very Serious Business': Managerial Relationships on the Ball Plantations, 1800–1835," *South Carolina Historical Magazine* 88 (January 1987): 3–4; for information on Carolina bays, see: Richard D. Porcher and Douglas A. Rayner, *A Guide to the Wildflowers of South Carolina* (Columbia: University of South Carolina Press, 2001), 43–45.

grew from 40 to 136 people.[42] Isaac Ball intensified his rice efforts in 1816 by increasing his enslaved population by forty-eight people. With more people working in the inland rice fields, the crop grew from 28.5 barrels of clean rice in 1816 to 159 barrels of clean rice in 1820, with production declining in 1823 to 129 barrels of clean rice.[43] Despite 433 acres of swamp at Jericho available for inland rice production, Isaac cultivated only 100 acres of rice in 1821. The next year, he increased his rice acreage to 129 acres. By 1823, Jericho slaves devoted 133 acres to rice cultivation. Despite this investment of labor and acreage, the unstable control of water decreased yields from 35.9 bushels per acre in 1821 to 19.5 bushels per acre in 1823.[44] In three seasons, Isaac Ball lost forty-three acres to freshets, with half of his loss from the "great hurricane" of 1822. Even with increased slave labor and greater acreage devoted to Jericho rice cultivation from 1816 to 1823, Isaac Ball's returns diminished steadily from 1.5 barrels per acre to less than a barrel per acre over that period. To put this into context, agricultural historian Lewis Grey wrote that a productive inland rice field could produce 2.5 to 3 barrels per acre.[45] Declining yields were a direct reflection of decreasing soil fertility in combination with natural disasters and variable labor output.

After Isaac Ball's death in 1825, Jericho once again went back to a secondary status. Isaac's brother, John Ball, Jr., managed the estate for the Ball family. After John Ball, Jr.'s death in 1835, Isaac's wife, Eliza Poyas Ball, and her brother, John Poyas, assumed the responsibility of managing the estate. Mathurin Gibbs came to Jericho Plantation through the Poyas connection, as Mathurin's wife Maria was Eliza's niece and John's daughter. Upon viewing the property before moving to the

[42] "Slave Lists, 1783–1843," John Ball Memorandum, "Isaac Ball Memo Book, 1821–1824," Ball Family Papers, SCHS; Cody, "Slave Demography and Family Formation," 63, 103 nn. 32 and 33; "Tax Return, Isaac Ball, 1824," Ball Family Papers, SCHS. For analysis on the South Carolina rice market economy, see: Henry C. Dethloff, A History of the American Rice Industry, 1685–1985 (College Station: Texas A&M University Press, 1988), 6–46.

[43] Ball and Gilchrist Papers, MS vol. 16, 1804–1890, USC; "Isaac Ball Memo Book, 1821–1824," Ball Family Papers, SCHS.

[44] "Tax Return, Isaac Ball, 1824," Ball Family Papers, SCHS; "Slave Lists, 1783–1843," John Ball Memorandum, Ball Family Papers, SCHS; "Isaac Ball Memo Book, 1821–1824," Ball Family Papers, SCHS; Cody, "Slave Demography and Family Formation," 63, 103 nn. 32 and 33.

[45] Gray, Agriculture in the Southern United States, vol. 2, 730; "Isaac Ball Memo Book, 1821–1824," Ball Family Papers, SCHS; for information on the hurricane of 1822, see Walter J. Fraser, Jr., Lowcountry Hurricanes: Three Centuries of Storms at Sea and Ashore (Athens: University of Georgia Press, 2006), 74–85.

plantation, Gibbs observed that the Jericho slaves were still "preparing the land for cultivation under the direction of the overseer," and, optimistically, he saw that "the place is in good order and more of the land prepared for cultivation than I can plant with my force."[46] Realistically, the property uninhabited by the Ball family for decades, had a less promising infrastructure than Gibbs portrayed. Mathurin's daughter noted later in her memoirs, how the main house was "shabby" and "not even lived [in]" for some time, with no glass in the window panes, and deteriorating conditions.[47] Her father ordered the house restored during the first year he was there, but he had problems with neglect in the rice fields for the rest of his life.

Gibbs' problems at Jericho revolved around cultivating a labor-intensive crop with a limited number of people. Labor problems began as soon as Gibbs moved his family to Jericho. Gibbs believed his enslaved Rice Hope laborers did not have adequate knowledge to grow rice on a large scale. Eighteenth-century Rice Hope slaves grew rice on a commercial level, but Gibbs' slaves grew only five to eight bushels of rice annually. After moving to Jericho, Gibbs integrated his slaves into the Ball workforce "still on the place." After two weeks, Gibbs realized correctly that his Rice Hope slaves were "unequipped with the cultivation of rice" and not knowledgeable enough to handle the new agricultural environment.[48] To counter this problem, Gibbs decided to sell eight Rice Hope slaves in the upcoming Ball estate sale and buy people who were working on the property. "Old Billy" became an important person at Jericho for his ability to revive rice cultivation on the property. Billy was a Jericho field hand whose ownership had transferred with the estate from one Ball heir to another. He knew the landscape and the art of rice cultivation. When the Ball estate sold the former Jericho driver, Simon, in January 1845, Billy, being of old age and of little monetary value at auction, was "promoted" to driver.[49] After assessing his purchases from the Ball estate sale, Gibbs expressed his frustration in that he sold off eight slaves of his

[46] 23 December 1844, Gibbs Plantation Register, SCHS.

[47] Mary Gibbs Ball, Ball and Gilchrist Papers, vol. 43, 22, USC.

[48] 23 December 1844, 3 January 1845, Gibbs Plantation Register, SCHS.

[49] Simon noted as driver in "Cloth at Jericho, 21 December 1840 to 21 November 1844." Billy was first called "Old Billy" in the Jericho cloth inventory 23 November 1843. Prior to that date, he was just called "Billy." Ball and Gilchrist Papers, vol. 16, USC; Isaac Ball estate sale advertised as "A prime Gang of one hundred and fifty NEGROES, belonging to the estate of Isaac Ball, accustomed to the culture of Rice on the Cooper River, among them are many carpenters, coopers, & c.," *Charleston Mercury*, 9 January 1845: 3.

own measured at four and a half full hands and bought nine slaves measured at two full hands. Despite the decline in perceived output, Gibbs believed this purchase was necessary because the Ball slaves knew how to grow rice. Summarizing his financial difficulties, Gibbs wrote, "it really seems as if ill luck has become enamored of us, and every change which we make to better our situation lends to increasing difficulties."[50] The average price Gibbs paid per slave was $350; Old Billy's price was $50. Gibbs lamented his purchase, noting that Billy was too old to work, by which he meant handling taxing labor. But soon Gibbs learned that Billy's value came from his knowledge of rice cultivation and serving as a "skillful manager."[51]

Rice historian James Clifton explains that often rice planters did not have detailed knowledge of rice culture; they had to "rely heavily" on either their overseers or drivers for the crop's success.[52] Judging from Gibbs' lack of output at Rice Hope, an inland plantation manipulated specifically for rice cultivation, Gibbs did not have even basic knowledge of the agricultural schedule; he admitted he must "rely on our [his] negroes for the crop."[53] At Jericho, Gibbs' limited capital prevented him from hiring an overseer, which would have cost between $250 and $600 annually. Gibbs' detailed notes of the cultivation cycle after Old Billy's arrival indicated the African-American driver had at least substantial knowledge of the process and, at best, was a critical factor in the crop's success at Jericho. According to Clifton, the driver knew all aspects of rice cultivation and supervised day-to-day operations. Billy's years working Ball rice fields had provided experience as to timing, field preparation, sowing seeds, flooding, and harvesting.[54]

Although Gibbs recorded which slaves performed which specific tasks, he mentioned Billy only rarely. With such a limited labor force, where every hand was needed to perform multiple daily tasks, one would expect Gibbs to discuss Billy working in the fields or even being absent. And while Gibbs took notes and omitted Billy in daily descriptions, it was this driver who directed the enslaved laborers to perform specific tasks.

[50] Isaac Ball, "A prime Gang of one hundred and fifty NEGROES," *Charleston Mercury*, 9 January 1845: 14 January 1845, Gibbs Plantation Register, SCHS.

[51] 14 January, 1845, Gibbs Plantation Register, SCHS.

[52] Clifton, "Rice Driver," 335.

[53] 31 December 1846, Gibbs Plantation Register, SCHS.

[54] Schantz, "A Very Serious Business," 7 n. 37; Clifton, "Rice Driver," 336, 338; Philip D. Morgan, *Slave Counterpoint: Black Culture in the Eighteenth–Century Chesapeake and Lowcountry* (Chapel Hill: University of North Carolina Press, 1998), 218–225, 342–346; Edelson, *Plantation Enterprise*, 79.

Frederic Law Olmstead made the observation while touring the antebellum South, "having generally had long experience on the plantation, the advice of the drivers is commonly taken in nearly all administration, and frequently they are, de facto, the managers."[55] With a brief journal entry, Gibbs provided a faded snapshot into this activity. Gibbs described how his cows wandered "into Old Billy's rice adjoining the new ground," also called Jim New Grounds, "and the only thing that saved that field from their depredations was the depth of water in the canal that separates that field from the one in which they got."[56]

Once again, a detailed understanding of Gibbs' rice output reveals the tedious nature of the Jericho microenvironment. Rice output varied during the four and a half years that Gibbs and his family lived on the property. Weather patterns had a direct effect on the crop's output, as freshets hindered growth during the first two years. In 1845, only 270 bushels were produced from sixteen and a half acres of rice fields, averaging just under 16.5 bushels per acre. A year later, Gibbs expanded his acreage by two acres, but yielded only 125 bushels or 6.76 bushels per acre, because of heavy flooding. By 1847, Gibbs' acreage and output increased. He expanded his rice fields to twenty-two and a half acres and yielded 360 bushels, averaging 16 bushels an acre. A year later, he became more ambitious in his field output, which produced 460 bushels, or 20.4 bushels an acre, but Gibbs became sick from malaria during the second half of the year. Complications from this illness would kill him the next year, leaving the plantation duties to his widow and oldest son. Even in his best crop year, Gibbs' average rice output per acre was less than other planters. Gibbs' best year of 20.4 bushels an acre barely reached the colonial average of 20 to 40 bushels per acre, while successful antebellum rice plantations averaged 30 to 50 bushels per acre.[57]

Even though Gibbs increased cultivated acreage and rice yields, his output remained below his peers down river and less than Jericho's

[55] Olmstead quoted in Robert William Fogel and Stanley L. Engerman, *Time on the Cross: The Economics of American Negro Slavery*, vol. 1 (Boston: Little, Brown, 1974), 213–214.

[56] 3 August 1847, Gibbs Plantation Register, SCHS; for drivers' privilege, see: Dusinberre, *Them Dark Days*, 124–125, 194–195; slaves were also growing rice at Limerick, who sold barrels to Isaac Ball between 1821 and 1825. David W. Babson, "Tanner Road Settlement: The Archaeology of Racism on Limerick Plantation," (M.A. thesis: University of South Carolina, 1987), 63.

[57] 3 June 1847, 7 December 1847, 31 Dec 1848, Gibbs Plantation Register, SCHS; Gray, *Agriculture in the Southern United States*, vol. 1, 284 and vol. II, 730.

output during Isaac Ball's tenure. For example, Gibbs' total cultivated rice acreage of twenty-two and a half acres in 1847 was dwarfed by Comingtee's 257 rice acres the previous year.[58] In comparison, Gibbs' acreage was less than a fifth of Isaac Ball's 133 cultivated rice acres in 1823.[59]

Controlling water on and off the rice fields was a constant struggle for Gibbs. Adequate enslaved labor to maintain embankments and cultivate rice fields was critical to Jericho. The Nicholson Creek placed heavy stress on Jericho rice banks with freshets flowing directly from Hell Hole Swamp. Storms consistently breached the reservoir ("Sand Dam") as well as individual field embankments. During two seasonal wet years, 1846 and 1848, slaves repaired breaks in the reservoir dam eight times and no less than five times, respectfully. Other embankments required mending, as breaks occurred throughout the fields during every significant rain recorded in Gibbs' ledger.[60]

Barely maintaining the rice field infrastructure, Gibbs did not have the necessary labor force needed for market production. After suffering foreclosure at Rice Hope, Gibbs relied on a marginal gang of twenty enslaved people at Jericho, out of which eight consistently worked in the fields.[61] He borrowed 18.5 bushels of seed rice from William J. Ball at Limerick and Quimby to start growing rice in 1845, yet he had to send Old Billy and Adam to Ball's Halidon Hill plantation eight days later to ask for additional seed. Optimistically Gibbs assigned one hand to an acre for task, but realized he was overextending his work force. The task of one person per acre to turn over soil and burn existing weeds at the beginning of the season, lessened to one person per three-quarters of an acre for more vigorous labors, like trenching drainage canals in and around the fields.[62]

A planter, like Gibbs, with limited labor, had to balance duties depending on prevailing needs and the agricultural calendar. There needed to be a division and direction of a plantation's enslaved labor force, to accomplish specific and cyclic tasks. Gibbs shuttled labor

[58] "Rice Land in Cultivation on Cooper River in 1846," Agricultural Experiments at South Mulberry, Sandford William Barker Papers, SCHS.

[59] "Isaac Ball Memo Book, 1821–1824," John Ball Memorandum, SCHS.

[60] 5 January, 31 January, 23 February, 25 February, 2 March, 9 March, 19 May, 13 June, 1846; 12 January, 15 January, 17 January, 24 January, 22 February, 1848, Gibbs Plantation Register, SCHS. Gibbs' ledger stops on April 26, so more breaks could have occurred in the rest of the 1848 season.

[61] Ibid., 14 January, 1 April, 1845.

[62] Ibid., 1–10 February, 9 April, 8 April, 11 April, 19 April, 1845.

between agricultural duties like planting rice, corn, potatoes, and other provisions for the plantation, and planting and monitoring conditions for the cash crop. During the four and a half years that Gibbs directed his enslaved laborers on Jericho, a stable routine evolved. When the New Year arrived, the slaves would continue to process remaining rice from the previous year. Gibbs' limited labor force could not tackle all the fields at once, so in January, three or four women turned over one half-acre of rice field soil and one or two men cleared one half-acre of irrigation trenches and canals, per task. Systematically, Gibbs assigned the tasks to one field at a time, and he planted the fields in the order of assignment. Heavy rains delayed completion of the task system. In 1846, Gibbs' rice cultivators were thwarted in their attempts to turn fields and clear canals between periods of rain. Although the gangs started their tasks in January of that year, rainstorms postponed work until the first week in February, and then again until the second week in April.[63]

The responsibilities of turning soil, burning debris, cleaning and retrenching irrigation waterways, and mending dams and trunks were divided among the slaves in January and February. By mid-February, other tasks directed the labor force away from rice culture. Instead they were instructed to cut wood, mend fences, repair buildings, prepare cornfields, and plant subsistence gardens. Gibbs pulled off some of the men to mend breaks in reservoir dams or field embankments, and this tied up his limited labor force for up to a week. Occasionally, fires on neighboring plantations encroached upon Jericho. While these fires never destroyed Gibb's crop, he did lose fence posts to the unconstrained flames. In March 1847, a severe fire burned a significant portion of Gibbs' fence that separated grazing lands from his Nicholson cornfield. Slaves spent several days putting out the fire and then had to repair "about 100 panels of fence," pushing back further the agricultural schedule.[64] Breaches in the fence allowed cattle to graze in the rice fields, destroying 22 percent of acreage in October.[65]

Labor resumed on the Jericho inland rice fields by the third week of March, after slaves had planted the cornfields. Planters attempted to have their rice fields sown as early as possible, yet to avoid an early spring frost; this enabled some planters to turn their fields twice a year. Gibbs, however, did not succeed in this endeavor. Slaves began sowing rice in April during only one out of the five years. The other four years, Gibbs' was

[63] Ibid., 6–16 January, 5–6 February, 10 April, 15 April, 1846.

[64] Ibid., 14 January, 20 March, 1847. [65] Ibid., 20 October 1847.

behind his ideal schedule, which would have had his field hands sowing rice from the third week of April to mid-May. Jericho slaves worked one field at a time, varying the amount of effort dedicated to total acreage, depending on weather and laborers' health. First, slaves chopped the turned soil with hoes, which loosened the earth and aerated the ground. Traditionally women performed this task, while the men followed behind them to lay out staked guidelines to designate rows and trenches. At the same time, other men cleared the drains of accumulated debris to enable efficient water control on and off the fields. The amount of land worked, depending on the variables described, wavered between one acre and one and three-quarter acres, for six hands.[66] Once a field was staked, women followed behind the men and planted rice seeds in the rows; they then smoothed the soil for aesthetic uniformity and efficient water flow.

Gibbs' lack of the labor caused him to run behind the agricultural schedule. Originally, Gibbs segregated his labor force by gender, assigning men to dig trenches while women sowed the rice plants. This practice was standard for planters, as Judith Carney explains in *Black Rice*. Even after the fields were sown, Gibbs forced the enslaved men back to clearing drains, while one slave managed flooded Jim New Grounds and another sought to keep the rice birds away from the newly sown crop.[67] By May 5, 1845, Gibbs was on a tight schedule to resume the entire practice again in the second field, "Pipkin." This field was named after Lewis Pipkin, a Jericho overseer in the 1820s, who was related to overseers at Limerick and Quimby.[68] While Gibbs intended to grow eleven and a half acres on Pipkin, he ran out of time and his slaves could complete only eight acres before flooding the field. Three days later he pulled his work force off Pipkin to begin hoeing out weeds at Jim New Grounds.[69]

Water management was the second problem Gibbs faced at Jericho. Growing rice in close proximity to Hell Hole Swamp and Nicholson Creek tributaries required both a keen understanding of hydrology and the ability to direct a large labor force in the successful management rice fields. Unfortunately, Gibbs did not possess either of these qualities. With

[66] Judith Carney, *Black Rice: The African Origins of Rice Cultivation in the Americas* (Cambridge, MA: Harvard University Press, 2001), 118–119.

[67] 17 April, 19 April, 1845, Gibbs Plantation Register, SCHS; Carney, *Black Rice*, 110–111, 118–119.

[68] Ball and Gilchrist Papers, MS vol. 16, 1804–1890, USC; Schantz, "Very Serious Business," 8–9, 11, 21.

[69] 20 May, 23 May 1845, Gibbs Plantation Register, SCHS.

streams flowing into his property from multiple directions, Gibbs battled for water control constantly. His problems multiplied when embankments leaked water or blew out. The first May that Gibbs lived at Jericho, a week's worth of rain overflowed the fields and embankments. He braced for the freshet: "Hell Hole is sending down its waters, and looks in the distance like a cloud of mist."[70] When the Pipkin dam leaked from too much water pressure, Gibbs ordered the trunk minder to release water through the upper field even though the rice had not started to sprout. There was then an overabundance of water and Gibbs "ordered water to be run off [Pipkin field] so as to let the last sown rice come up, and when it has sufficient strength to flow Jim New Grounds." He was forced to balance the structural integrity of the embankment with the schedule of the rice plant.[71]

Constant rain-induced flooding in 1846 prevented Gibbs' field hands from sowing rice for two weeks. Eventually the storm took its toll on a weakened embankment. "The rain yesterday prevented the continuance of the labors of the rice fields," explained Gibbs. "One of the banks [has] broken a chasm of about 15 feet," which one of the hands set out to repair. Gibbs voiced his frustration again a month later over lack of progress in the rice fields. "It is unfortunate to those who have to sow rice at this late in the season," he wrote in his journal, "for little can be done and the rice fields have been so full during the spring as to prevent early planting in inland situation, where getting off the water is very great."[72] Hell Hole Swamp water flow proved too much for Gibbs. He noted, the reservoir dam overflowed and made the land "swampy" where the Nicholson Creek "branches are full to overflowing. It still continues to rain and all of Nature is dripping and uncomfortable."[73] The unfortunate situation limited Gibbs' output to less than half that of the previous year. Only 125 bushels came from the crop in 1846, once more limiting the capital Gibbs needed to pay debts and purchase more labor.[74]

The combination of limited labor with which to grow rice, leaking embankments, and heavy downpours thwarted Gibbs' agricultural cycle and forced him to rearrange his task schedule. Spring rainstorms

[70] Ibid., 20 January 1845.

[71] Ibid., 15 May, 17 May, 1845; [Robert F.W. Allston], "The Rice Plant," *DeBow's Review* 2 (March 1852): 296; 28 May 1845, Gibbs Plantation Register, SCHS.

[72] 9 May 1846, Gibbs Plantation Register, SCHS; 8 June 1846, Gibbs Journal, Ball and Gilchrist Family Papers, USC.

[73] 11 June 1846, Gibbs Journal, Ball and Gilchrist Family Papers, USC.

[74] 31 December 1846, Gibbs Plantation Register, SCHS.

prevented rice fields from drying, which in turn prevented his slaves from pulling weeds. At least once a year Gibbs had to delay field hoeing because of rain, pushing back the list of tasks needed to complete a profitable season.

For inland rice planters with a limited labor force, disruptions to their agricultural schedule caused a ripple effect for the rest of the season. In Gibbs' case, he had to make adjustments to his embankments breached by water. In 1847, a late spring rain weakened the floodgate at Jim New Grounds. Male hands spent two days attempting to replace the leaking gate, but were unsuccessful in holding back the flow. Gibbs made the decision to dam up the floodgate and trunk, sealing the Jim New Grounds embankment entirely, to prevent further unwanted flooding of the field. Ironically, the two men assigned to seal the embankment were rained out.[75] After ten days of his enslaved weeding Pipkin Field, Gibbs resumed the efforts of mending the New Ground dam, which he feared would not hold.[76]

Constant July rain disrupted Gibbs' work schedule even further. Gibbs had to force his laborers to hoe New Ground and Pipkin, between the long and the harvest flows, in the rain. The increasing rainfall by the end of the month, however, put a stop to this task; the storms "put the rice under water," and prevented any hoeing from occurring. Three days later, Gibbs commented that, the "rice fields are too full of water to be worked." Even with standing water in these fields between scheduled flooding, rainfall would back up the waterways, preventing people from flowing water out of the fields. A month of summer storms created a critical mass of water flow through the Jericho Swamp and Nicholson Creek, which prevented canal drainage. Gibbs' accepted his inability to force people to work in the fields and redirected his laborers to spend the first week of August mending fences where the cattle had encroached upon the rice fields.[77]

With two months of rain disrupting the agricultural schedule, Gibbs began a two-week push on August 10 to hoe the fields quickly before the final flow. He sent a gang down to New Ground and the lower part of Pipkin field; the slaves "could only reach the flood gate at Abram's Corner where the dam is broken."[78] Gibbs, sensing the pressure to finish, increased the hoeing task from one acre to two acres with four hands.

[75] Ibid., 4 June–16 June, 1847. [76] Ibid., 26 June 1847.
[77] Ibid., 13–23 July 1847; quote: 27 July 1847; quote: 30 July 1847; 3–9 August 1847.
[78] Ibid., 10 August 1847.

By the end of the month his laborers finished hoeing the fields. In addition to preventing cows coming into the fields and weeding a heavy concentration of grass in Barn Square, they accomplished one acre a day per hand. The unusually wet season sent a ripple effect through Gibbs' cultivation schedule. The heavy rains in April and May created delays in the agricultural cycle, daunted efforts to mend breaks to achieve flood control, and finally, imposed delays in hoeing because of the high weed content. As a consequence, Gibbs had to stagger his tasks and limit his labor output. He realized he was behind schedule compared to his peers. When he was at the end of the final hoeing, other inland planters were two to three weeks into their final flow.[79] Flood problems continued near the end of the season. Barn Square had not reached its final flow stage because grasses and cattails contaminated the fields. To eliminate these weeds, four women were told once again to increase their workload, this time to two acres per hand per task. However, it took these women one week to clear four acres, as they had to hoe in the water, pick cattails, and remove volunteer rice.[80] At the end of the season, Gibbs noted new problems with flooding and pests that "shed upon me[,] in this part of my crop[,] anxiety for the safety of this crop." On the final day of the year Gibbs reflected how despite rain, "prevent[ing] vigorous growth," and cattle, birds, and "wild animals" destroying the "provision crop, on which a planter must rely for the food of his plantation people as well as stock," he stated "the rice crop has exceeded all that I have produced since I have been a planter."[81] To Gibbs, cultivating enough crop for the next season, amidst all the natural disasters, was a blessing indeed.

Land three miles downstream from Jericho also presented unique challenges in terms of planting, water control, and output. Gibbs neighbor, Dr. John Beaufain Irving, owned another reservoir-fed rice plantation called Windsor. Irving, like Gibbs, came from the elite planter social circle without direct inheritance. Irving's great grandfather, James Irving, was born in Scotland, earned a degree in medicine and moved to South Carolina in 1745; he married Elizabeth Motte, daughter of Jacob Motte, public treasurer of the providence, who at the time held the most profitable office in the provincial government. James Irving established connections in Charleston through marriage, profession, and his purchase of

[79] Ibid., 27 August 1847; [Allston], "The Rice Plant," 296–297; Cody, "Slave Demography and Family Formation," 107.

[80] 14–21 September 1847, Gibbs Plantation Register, SCHS.

[81] Ibid., 31 December 1847.

Howe Hall, a 1,470-acre inland rice plantation located on a tributary of Back River, in St. James, Goose Creek, from Thomas Middleton on November 15, 1755. Aggressively increasing his plantation capital, James traded this property within three weeks for land and slaves in Ironshore, St. James, Jamaica. The family moved to this Caribbean island the same year and created a successful sugar plantation, which their descendants maintained up to the twentieth century. Despite relocating to Jamaica, the Irvings maintained close relations with Charleston society. James' son (father of John Beaufain Irving) also married into Charleston privilege. Hannah Margaret Corbett was the daughter of Thomas and Margaret Corbett, a Charleston merchant and Cooper River plantation owner. Also, Hannah was the granddaughter of Captain John Harleston, who owned Irishtown Plantation on Turkey Creek, the second tributary forming Huger Creek and five miles southeast of Windsor.[82]

John Beaufain Irving represented the upper echelon of James Oakes' "middle-class masters," transitioning from professional to planter.[83] Irving was born into privilege at Ironshore on September 28, 1800. While his parents managed the Jamaican plantation, John spent his early, formative years in the South Carolina Lowcountry. Between four years old and ten years old, he lived either in Charleston or out on his grandmother's Farmfield plantation, located on the East Branch of the Cooper River between the Ball family compounds of Comingtee and Kensington. Upon graduation from Cambridge University, Irving returned to the United States to study medicine in Philadelphia; he moved to Charleston by 1823. John Beaufain Irving, being the third son, did not inherit any right to his parents' Jamaican property. Nonetheless, with an upbringing steeped in planter aristocracy, Irving was able over time to acquire capital and maintain connections with the landholding gentry. Here Irving straddled between maintaining a reputable occupation while jockeying to maintain elite social standing among his peers in the landed gentry. Irving grew his medical practice in Charleston for seventeen years, before purchasing Windsor in November 1840 for $2,000. By the time Irving

[82] Dr. John Beaufain Irving Family Papers, folder 30-4, SCHS; M. Eugene Sirmans, *Colonial South Carolina: A Political History, 1663–1763* (Chapel Hill: University of North Carolina Press, 1966), 252–254; Henry A.M. Smith, "Baronies of South Carolina: Quimby and the Eastern Branch of the Cooper River," *South Carolina Historical and Genealogical Magazine* 18 (January 1917): 14–15.

[83] Oakes, *Ruling Race*, 57–65.

secured the 1,150-acre property from Catharine Edwards' estate, the plantation was in disrepair and not producing rice.[84]

Windsor was subject to uncontrollable freshets. Irving's agricultural journal revealed its land was disposed to freshets because the tight floodplain corridor was nestled between two moderately declining landforms. Like those of Gibbs, Irving's descriptions of water overflowing the reserve dam and then rushing sequentially into impounded rice fields, occurred on an annual basis. Both Irving and Gibbs observed closely how their embankments were holding up, when a freshet backed up their reservoirs. Each planter hoped not to discover a breach that would send a wall of water to the adjoining fields. However, an October 1845 freshet had polarizing effects for Gibbs and Irving. Three days of rain resulted in destructive floods. For Gibbs, the freshet blew out his lower embankment at Jim New Grounds; the storm weakened two more breaks later that month. For Irving, the same freshet brought minimal destruction, and "the only [damage] sustained was washing away a little of the top Earth on my Back dam."[85]

Several factors explain why Irving and Gibbs did not suffer similar consequences in the 1845 freshet. The first was the construction integrity of the embankments. Possibly Gibbs' embankments were not as firmly packed by his enslaved laborers; or the soil used on these embankments proved too sandy, with not enough clay to coalesce the alluvium; or perhaps the seemingly constant freshets, directly flowing downstream without any human barriers, put too much strain on Jericho's earthen structures. The two plantations had somewhat consistent soil patterns, each with the Meggett series in stream and creek floodplains and the Bethera series, combined with minor soils, in the higher elevations. Meggett loam is found in well-defined drainage-ways and consists of a poorly drained, loamy surface with clayey subsoil. Bethera loam is similar in composition, with a loam surface and clayey subsurface, yet has a higher percentage of sand. Jericho's floodplain, broader than that of Windsor, incorporated water from multiple tributaries of the expansive Hell Hole Swamp and channeled large amounts of water into Nicholson Creek.

[84] Dr. John Beaufain Irving Family Papers, folder 30-4, SCHS; Aemiluis Irving, and L. Homfray Irving, ed., *James Irving of Ironshore and His Descendants, 1713–1918*, (Toronto, ON: College Press Ltd., 1918), 107–108; J.B. Irving, "Record of Windsor and Kensington Plantations, 1840–1888," 1840, CLS.

[85] 4 November, 28 November, 1845, Gibbs Plantation Register, SCHS; 21 October 1845, "Record of Windsor and Kensington Plantations," CLS.

Windsor received water from Nicholson Creek to form a single passage through the rice fields.[86]

Windsor's water flow, with less strain on embankments and the reserve dam, was the result of several generations of landscape alteration. Between 1788 and 1840, Evan Edwards and his wife, Catharine, and family shifted water management on the plantation. A 1788 plat of Windsor shows four rice fields, with no reservoir on the property. Edwards purchased adjoining Nicholson Plantation in 1793 to acquire more land along Nicholson Creek; this acquisition of 555 acres served as a reservoir site and provided additional rice fields. Some time during Edwards' occupation of Windsor, an upper field became a reservoir. Later owner Irving referred to what had been the division between the third and fourth fields as the "Back Dam."[87] It was common for neighboring inland planters to use this label, as the reservoir impoundment was significantly higher and wider than the average field embankments. The Back Dam exists today, holding water to form an impounded wetland and green tree reservoir that maintains a high concentration of cypress and hardwoods.[88]

The Back Dam turned a 100-acre field into a sprawling reservoir, located strategically at the broadest section of Nicholson Creek floodplain. The Edwards family sacrificed valuable agricultural land to create this reservoir. One can still see cross banks stretching across the former field today; a rice field swapped for a line of defense from detrimental natural disasters. Perhaps Edwards could not keep freshets out of this upper field, or possibly he expanded his reservoir in times of drought, rearranging the function of the field. Regardless of his reasons, Edwards' rearrangement of the upper division placed higher volumes of impounded water within Windsor's borders to flood downstream fields.

The "Tanner Fields" represent how people marginalized by plantation society participated in the inland rice cultivation process. The field's namesake, Edward Tanner, was a mulatto freeman who lived on a settlement in the boundary lines of Limerick, and the northwest corner

[86] *Soil Survey of Berkeley County*, 5–6, 8–9, 22–23, map sheet no. 58.

[87] 11 January, 4 August, 21 October 1845, "Record of Windsor and Kensington Plantations," CLS.

[88] "Back Dam" is also referred to as Midway's reserve dam, 6 April 1830, Ball Family Papers, SCHS; 2 January 1850, Keating Simons Ball Papers, UNC; also, back dam located at War Hall, a 1715 inland rice plantation near the Edisto River, St. Paul's Parish, 8 February 1771, Taylor Family Papers, USC. "Thomas Osborn to Evan Edwards, 2 December 1793," Deed Book M6:405, CCRMC.

of Windsor. He was born on Comingtee Plantation to Dolly, a slave, and if Edward Ball's speculation in *Slaves in the Family* is true, Tanner's father was Elias "Red Cap" Ball. To avoid criticism from neighboring planters and family members, Elias arranged for the relocation of Dolly and newborn Edward immediately; they moved up the East Branch of the Cooper River to Quimby Plantation in 1740 and later to the remote St. Johns Plantation, where they lived until Elias Ball's death in 1751. Edward received his freedom after his father's death and by 1763 moved to Kensington Plantation, which was owned by his half brother, Elias the second. Once at Kensington, Edward established a career as a leather tanner. He produced income by making and selling slave shoes, harnesses, and saddles to neighboring plantations. In addition, he provided medical attention to slaves. By 1790, Tanner moved to the eastern boundary of Elias II's neighboring plantation, Limerick. He called this settlement, "King Robin" and lived there until his death in 1820.[89]

Using King Robin as a central location between Limerick and Windsor, Tanner established business connections within an eight-mile radius. Besides manufacturing a variety of leather goods for the Balls, he sold thirty-eight pairs of "negro" shoes to John E. Poyas at Richmond Plantation and twenty-nine pairs of shoes to Edward Harleston at Fishpond Plantation. Both Richmond and Fishpond are located on the East Branch of the Cooper. Tanner sold 285 pairs of slave shoes over a two-year period to Major Isaac Harleston of Irishtown Plantation. Also, Tanner provided medical services to George, a slave at Windsor, and sold hides and made saddle repairs for Robert Quash at Fishbrook Plantation on Turkey Creek. Providing specialized products and services to the Huger Creek area plantations enabled Tanner to continue building capital and purchasing more slaves. Tanner became a slave owner by 1790 with three enslaved people under his direction. Within ten years he owned seven people, and by 1820, the number grew to nine.[90]

Tanner's business transactions illustrate how his connections with neighboring planters enhanced trust between and promoted fluidity of agricultural exchange. Tanner's settlement sat on high ground next to a tributary that fed into the area of Tanner Fields, a division of the Windsor

89 Ball, Slaves in the Family, 187–189; "Estate of Ed Tanner Account Book," John Ball, Sr., and Jr., Papers, Duke University (DU).

90 Ball, Slaves in the Family, 187–189; Babson, "Tanner Road Settlement;" "Estate of Ed Tanner Account Book," John Ball, Sr., and Jr., Papers, DU; Ball and Gilchrist Papers, MS vol. 15, 1783–1810, USC; Morgan, Slave Counterpoint, 487.

rice fields. By 1800, Tanner owned seven slaves; it is not clear whether or not Tanner's slaves cultivated rice at the nearby Windsor fields. It is possible Catharine Edwards diversified her income by renting one of the rice field divisions to Tanner. Isaac Ball referred to Tanner Fields in his Limerick agricultural journal; Ball's slaves rotated rice, oats, and cotton in that field. Surviving receipts from neighboring inland planters (including Evan Edwards at Windsor, Isaac Harleston at Irishtown, and Robert Quash at Fishbrook) document only Edward Tanner's leather-craft. However, Tanner's death certificate listed him as a planter by 1820.[91]

Tanner's location also provides some context as to how this African-American slave owner lived in a "peripheral settlement": a tract sandwiched between two plantations and not having a direct connection with the central plantation settlements. Tanner's King Robin settlement lay on the Ball side of the boundary between Elias Ball's Limerick and Evan Edwards' Windsor. It was just off the main road dividing the two properties. Although there is no record of Tanner actually owning this settlement, one can speculate that Elias Ball leased the property to his half-brother, Edward Tanner. Tanner's upland settlement, although distant from the white social structure, was central to Tanner's business connections and landscapes. He accessed the navigable Huger Creek from Limerick's dock and traveled to the highway between the Cooper and Santee Rivers from the Limerick public road. Cattle used for leather grazed within his fenced property, as Tanner's family and enslaved labor lived and worked in three roadside buildings at King Robin. Edward Tanner's enslaved laborers worked in the Tanner Fields on the Windsor side by following the creek down hill less than a quarter of a mile.[92]

The presence of Tanner slaves on Windsor raises questions about crop output and also control of natural resources. It isn't clear whether Tanner's crop was sold at market, or whether it served as a part of the provisions for the settlement's families. It is also worth asking if Tanner worked with Catharine Edwards to dictate the amount of water pulled from the reservoir, and, in times of drought, if tensions arose between Tanner and Edwards over the amount of water Tanner could use in his rented field. From archaeological and historical evidence, Edward Tanner

[91] The National Map, http://nationalmap.gov/; "Will of Edward Tanner," 9 May 1820, "Estate of Ed Tanner Account Book," John Ball, Sr., and Jr., Papers, DU; 1819, 1823, 1825, vol. 16, "Limerick Plantation Register," Ball and Gilchrist Family Papers, USC.

[92] Babson, "Tanner Road Settlement," 49–51; "Plan of Limerick, a Plantation Belonging to Elias Ball, Esq.," June 1797, no. 1541, John McCrady Plat Collection, CCRMC.

created a comfortable life for his family. Isaac Ball's inventory of Edward Tanner's possessions totaled $3,722.75. Although rice was a usual avenue to wealth at the turn of the nineteenth century, it may have been a minor venture only for Tanner. Nonetheless, the elaborate collection of excavated pottery shards suggests Tanner was successful, with money in the family to purchase current European ceramics. At first, based on the artifact distribution pattern, archaeologists assumed in error that Tanner was Euro-American with African-American slaves. Tanner obtained the "planter" title by his death, and whether this was a title he sought and earned through inland rice cultivation, or one added as an afterthought on his death certificate, it remains that Tanner prevailed in using his association with inland rice cultivation as a mode to achieve a higher status.[93]

When John Beaufain Irving acquired Windsor, the landscape changed once again. Irving purchased the property from Catharine Edwards' estate in November 1840 and "found the place much out of order."[94] With no Windsor field hands the first two years, Irving did not have enough labor to grow agricultural commodities for profit. Then, in April 1842, Irving combined his seven slaves with gangs borrowed from Alfred Huger's Longwood Plantation and Dr. Benjamin Huger's Richmond Plantation, both located on the East Branch of the Cooper, to plant ten of the forty-five impounded acres on Fishbrook Field. The following year, with Alfred Huger's help, Irving planted eighteen acres on Fishbrook Field, "which as before yielded well." Summarizing his agricultural accomplishments in 1843, Irving wrote, with "no adequate force to carry on the improvements I contemplated when I first took possession of Windsor, I have made but little progress."[95]

Irving, like Mathurin Guerin Gibbs two years later, wrestled with the problem of not having enough labor to maintain a productive inland rice environment. Although Irving bought and borrowed slaves during his first three years at Windsor, the work force was not enough to suitably cultivate rice. Historian James Clifton notes that rice planting was profitable only "through quantity production" and thus planters maintained sizable labor populations.[96] Neither Irving nor Gibbs produced large quantities of rice when relying on their small number of enslaved people

[93] Babson, "Tanner Road Settlement," 72; "Will of Edward Tanner," 9 May 1820, "Estate of Ed Tanner Account Book," John Ball, Sr., and Jr., Papers, DU.

[94] November 1840, "Record of Windsor and Kensington Plantations," CLS.

[95] Ibid., 1843. [96] Clifton, "Rice Driver," 331.

to perform the tedious tasks of digging and clearing canals, turning and chopping fields, sowing rice seeds, hoeing and pulling weeds, and harvesting the crop. Additionally, these laborers had to quickly and successfully mend any breaks in embankments, plus repair damaged floodgates and trunks. Simply put, before 1844, Irving could not produce a profitable output from his inland environment.

The year before Gibbs moved to Jericho, Irving was optimistic that he would have enough slaves and knowledge to harvest a productive crop. The actual outcome dashed those hopes. After receiving a gift of fourteen slaves from his aunt, Elizabeth Corbett, in late 1843, and adding an additional eight enslaved laborers purchased from Sims White in February 1844, Irving declared, "this year I commenced a plan of internal improvements which I pursued with much energy." He devoted four months of his slaves' labor to strengthening Back Dam, setting new floodgates, clearing canals, and planting "the whole" Fishbrook field. Ironically, with all the energy directed toward water management, the rains never came in 1844. A "great drought" struck the region, and Irving was unsuccessful cultivating his crops.[97] In a report submitted to the State Agricultural Society, R.F.W. Allston stated, "the Planters on the Cooper River have suffered most severely – that river having been salt a considerable extent throughout the season." The report noted that the Cooper River planters produced only 12,000 barrels of rice for the 1844 season, where before they averaged 22,000 barrels. Although Allston's subjects were using tidal irrigation, Irving did not escape the natural disaster, stating, "almost all lost to the prevailing drought."[98]

To expand his labor force to a size capable of cultivating this inland rice environment, Irving purchased nineteen slaves from Isaac Ball's estate sale in January 1845. (The same estate sale where Mathurin Gibbs bought his enslaved force.) Some of the people came from Jericho and brought with them intimate knowledge of rice cultivation along the Nicholson Creek watershed. The same month, Irving integrated his enslaved labor force with free Irish laborers to dig a canal, which connected the second

97 1843, 1844, "Record of Windsor and Kensington Plantations," CLS.

98 [R.F.W. Allston], "Report on Rice," *Southern Cultivator* 3 (1 February 1845): 19; 1844, "Record of Windsor and Kensington Plantations," CLS. Charleston chemist J. Lawrence Smith presented a report to the Agricultural Society in 1845 as a direct response to the drought. Smith proposed that tidal rice planters marl their soil more frequently during extended droughts to suppress the salinity in the fields. J. Lawrence Smith, "Effect of Drought upon Rice Lands," *Southern Agriculturalist* 5 (November 1845): 425.

field division's drainage canal around Tanner's Field to the Back Dam's "wasteway."[99] The finished canal was eight feet wide and a half-mile long. Upon completion, the additional waterway provided a continuous flanking canal that bordered the three Windsor rice fields. A two-day storm in August 1845 put the new flanking canal to the test. Irving recorded with pride that the "water which accumulated in my Reserve which run off without any overflow of the adjoining fields," meaning that floodwater successfully flowed around the rice field divisions instead of breaching the field embankments.[100]

While inland rice planters fell victim to natural disasters more often than their tidal counterparts, there were occasions where the inland planters faired better than their tidal peers. For example, two successive years of drought beginning in 1845 compounded irrigation problems for Lowcountry tidal rice planters. Tidal rice planter J. Motte Alston called 1845 the "great salt-water year" based on brackish water contaminating the rice crop.[101] However, Irving was one of the fortunate rice planters who had "fresh" land resulting from inland reservoirs. With fifty-three slaves, plus a newly hired overseer, Irving embarked on planting fifty-three "Riceland" acres at Tanner Field and twenty-two acres of "Rice lands" in the adjoining field. By April, Irving braced for potential devastation to his crops. He saw that the drought was a problem for tidal planters along the Cooper, as "threatening as last year," with "salts" up to Huger's Bridge which is the line of demarcation of tidal flow on the East Branch of the Cooper. The brackish tides killed the newly sprouted crop during the "point flow," and led some planters to start anew by replanting their crop.[102]

Despite the challenges inland planters faced during droughts, Irving harnessed enough water to irrigate his 1845 crop. In May, Irving planted

[99] 10 January and 11 January 1845, "Record of Windsor and Kensington Plantations," CLS; William Carson also employed Irish laborers to trench and build embankments on his tidal- and reservoir-irrigated rice fields, located at Dean Hall, West Branch of the Cooper, February 1843; see Mathew, *Private Diary*, 63.

[100] 5 August 1845, "Record of Windsor and Kensington Plantations," CLS.

[101] Between June and September, rain fell on only 43 days in 1844 and 42 days in 1845. Cary J. Mock, "Documentary Records of Past Climate and Tropical Cyclones from the Southeastern United Sates," *PAGES News* 10 (December 2002): 21–22; Arney R. Childs, ed., *Rice Planter and Sportsman: The Recollections of J. Motte Alston, 1821–1909* (Columbia: University of South Carolina Press, 1953), 62; 20 May 1845, "Record of Windsor and Kensington Plantations," CLS.

[102] April 1845, "Record of Windsor and Kensington Plantations," CLS; 15 April 1845, Gibbs Journal, Ball and Gilchrist Family Papers, USC.

the additional twenty-two acres in Windsor's central field, feeling confident even in the middle of a severe drought, and noting he had "plenty of reserve water at my command," while downriver, tidal planters suffered from encroaching brackish water.[103] Occasional rain divided the summer drought. For Irving's fields, the rice grew "vigorous and beautiful" during early point flowing in May, began to "ear out" in July, and was "riping [sic] fast" by mid-August.[104] Irving's agricultural success during the 1845 drought revealed how this natural disaster did not affect all rice planters. Inland planters, through the use of large watersheds, bays, and springs, did not suffer the same dilemmas as tidal planters. Irving expressed his solidarity to their problems, but his reservoir-irrigated crop enabled him to plant the following year. He observed in late August that the drought was "extensive and unparalled [and] that not only will the Rice Crop fall short, but the people in the upper part of the state began to be alarmed at the falling off in their Corn, and think of no alternative left but to immigrate. Public meetings have been held in several Parishes to devise the best means of relieving the necessities of the Poor and the Needy." Ironically at the end of August, Irving's rice harvest was delayed one week because heavy rains prohibited his labor force from cutting the crop.[105]

The flanking canal helped Irving to flood his three fields systematically at different times. Irving did not have enough of a labor force to plant his entire rice crop simultaneously. As seen in 1845, Irving planted two fields of rice seven weeks apart.[106] With two fields on different cultivation schedules, the first field's flow was up to a month behind Tanner Field's. In a simple inland irrigation system, water flows from the reservoir, through the first field, and then down stream to successive fields. If there was a delay sowing the lower field after sowing the upper field, water traveled first through the sown upper field adjoining the reservoir and then continued to the lower fields. This early process of transferring water from one planted field to the next required skill to maintain the desired water height necessary for the mature crop while flowing enough water in the newly planted field down stream. By irrigating fields with the flanking canal, this irrigation balance of transferring water from one field to the next was not a dilemma. For Irving, whose flanking canal connected the reservoir to each of his three fields, flooding each of his field divisions did

[103] 20 May 1845, "Record of Windsor and Kensington Plantations," CLS.
[104] Ibid., 20 May, 14 July, 13 August, 1845. [105] Ibid., 20 August 1845.
[106] Ibid., 1 April, 20 May, 1845.

not impede on other field schedules. This flanking canal, while allowing freshets to bypass the fields, allowed too for the trunk minder to methodically flood or drain each of the fields without altering the progress in neighboring fields.

At the end of the 1845 season, John Beaufain Irving produced a good crop. His output for 1845 was just over thirty bushels of rice per acre compared to Mathurin Guerin Gibbs' output of just over sixteen bushels an acre for the same year. The first harvest yielded 1,674 bushels of rough rice, which he sent to a Charleston mill for pounding, threshing, and barreling. The second crop he sent downstream to the Middleburg Plantation mill, which yielded twenty-eight barrels of clean rice. The sustained drought and occasional freshets disrupted the 1845 Lowcountry rice cycle; the low output and increased demand created high prices during the fall 1845. The average price of rice was 3.5 cents/pound for that year, but the Charleston Market fetched up to 5.25 cents/pound in October 1845. Irving speculated that the price would rise even more, but that was not the case. By April 1846, when Irving decided to sell his clean rice, the price in Charleston had dropped to 3.99 cents/pound. He received $671.07 for his clean rice and sold the rough rice to the Charleston mill in May 1846 at 2.65 cents/pound for a total of $1997.67. After charges and processing fees, Irving made $2,255.90 for the 1845 crop.[107]

Following the 1845 season, Irving took assertive action and increased his labor force. He purchased two separate gangs of enslaved people, one with twelve people from Edward Harleston's Irishtown plantation and a second with eight people from Elias O. Ball's estate. An additional twenty people came to live on Windsor Plantation. The 1846 season, however, did not live up to Irving's expectations. Irving penned only one agricultural entry in his journal, revealing frustration and bemoaning the "unusually" wet season and that he could plant only twenty-five acres on Tanner Field.[108] As noted in Gibbs' agricultural journal, the 1846 season produced tremendous amounts of rain, inhibiting planting in these low-lying fields.

The same year, Irving redirected his agricultural pursuits by purchasing Kensington, a former Ball plantation located on the tidal banks of the East Branch of the Cooper River, for $6,900.[109] Irving may have considered the property a solid financial investment, as its fields relied on tidal

[107] Statement of Crop 1845, "Record of Windsor and Kensington Plantations," CLS.
[108] Ibid., 17 January, 3 March, bottom of statement for 28 March, 1846.
[109] "Mortgage of Plantation," 17 February 1846, Deed Book K11:424, CCRMC.

irrigation and represented the upper-echelon of South Carolina rice farming, or he may have been drawn to the acquisition because of the plantation's historical importance as an established Ball family tract since 1747. Irving used Windsor as a way to enter the planters' realm, and he managed the inland challenges to profit from Windsor's crop. Kensington Plantation represented his rising to the next level within the planter hierarchy. Involvement in tidal irrigation enhanced Irving's connections to his contemporary planters as he was practicing the same style of agriculture. The property was more visible to his peers – it was located along the navigable section of this prosperous branch of the Cooper River – and Barony Road passed through the property.

Irving's membership in agricultural and social societies symbolized his social stature through agricultural pursuits. Membership in the regional and state agricultural societies enabled social networking for, and sharing ideas amongst, the planter elite. Irving was a founding member of the Black Oak Agricultural Society, which in April 1842 consisted of St. John's Berkeley planters. Membership included French Huguenot descendents primarily residing in middle and upper Berkeley County, but did include notable planters along either branches of the Cooper River. Irving's membership lasted until April 1845. In 1847, Irving and his son, Aemilius, helped found the Strawberry Agricultural Society, an alternative to Black Oak; it attracted planters along the East Branch of the Cooper. With meetings held in the Strawberry Chapel parish house, on the grounds of the spiritual epicenter for the Ball, Harleston, and Laurens families, this society catered to a more intimate networking of social peers in this locale. Notably, Irving's maternal grandparents, Thomas and Margaret Corbett, were buried in the Strawberry cemetery. Irving was a consistent member of the society, as evidenced by the frequent inclusion of his name in their minutes. He became secretary and treasurer in 1852.[110]

Inland rice plantations provided a mode for merchants and professionals to enter into the elite planter class. Despite the discounted price of these tracts, planters had to manage water and crop carefully. Planters still had to depend upon a labor force to carry out water management and rice production. Planters' depended upon their slaves for skillful observation of how the fields worked within the immediate environment. Rice output was limited to availability of water and labor, without which the crop could

[110] "Black Oak Agricultural Society Minutes, 1842–1861," Thomas P. Ravenel Papers, SCHS; "Strawberry Agricultural Society Journal," Ball and Gilchrist Papers, vol. 21, USC; Mabel L. Webber, "Inscriptions from the Church-Yard at Strawberry Chapel," *South Carolina Historical and Genealogical Magazine* 21 (1920): 166.

easily perish due to an unforeseen natural disaster. The ability of planters to maintain these temperamental plantations, with the enslavement of people, was the cost of entering into an aspiring social status.

The stories of how Mathurin Gibbs and John Irving relied on inland rice cultivation as a mode to achieve financial success and upward social mobility had different outcomes. Gibbs, after contracting malaria in summer 1848, struggled for almost a year with "violent" colds and swelling of the face, and it hindered his walking until his death in May 1849.[111] Ultimately, the unhealthy landscape at Jericho, with its expansive wetlands providing breeding grounds for anopheles mosquitoes, killed Gibbs. In 1855, Mathurin's widow, Maria, and her youngest son Frederic moved to Windsor where they practiced rice cultivation anew on the former Irving property. Windsor's less isolated location and close proximity to Hyde Park Plantation, where the Gibbs' eldest daughter, Maria Louisa, lived with her husband John Ball, appealed to Maria Gibbs. The Gibbs and Ball family ties strengthened further when Mathurin and Maria's two younger daughters each married a Ball father and son. Mary Huger Gibbs married William James Ball, of Limerick, and Catherine Theus Gibbs married William James Ball, Jr., the oldest son from his father's first marriage to Julia Cart.[112] Within this inland nook along the Huger Creek, the Ball, Poyas and Gibbs families maintained their control of the land and over people through marriage and labor.

Irving, in comparison, sold Windsor and continued to plant at Kensington in the 1850s with some success, and eventually paid $26,666 for Farmfield Plantation in 1861. Despite what must have been an astounding price for a marginal rice planter like Irving, this property served as a sentimental retreat for him. Farmfield was where Irving lived with his grandparents, Thomas and Margaret Corbett, when he was between four and ten years of age. Irving and his son, Aemilius, continued to plant rice at the two tidal plantations until the Civil War, yet could not retain labor forces after emancipation. After his wife's death in 1867, Irving sold his properties and moved to Bergen, New Jersey. He moved near his eldest son, the painter John Beaufain Irving, Jr., and managed the New York Jockey Club until his death in 1881.[113]

111 Mary Gibbs Ball Memoirs, Ball and Gilchrist Papers, vol. 43, USC.

112 "Maria Gibbs to James Coward, mortgage," 1 January 1855, Deed Book I13:467, CCRMC; Deas, *Ball Family*, 146–148; Ball, *Slaves in the Family*, 202–203, 335–336.

113 Dr. John Beaufain Irving Family Papers, folder 30-4, SCHS.

6

Inland Rice Cultivation and the Promise of Agricultural Reform

Standing before the State Agricultural Society in Columbia, South Carolina, Henry William Ravenel, Secretary of the Black Oak Agricultural Society, presented the local society's driving research at the time. As part of the State Society's mission to feature regional agricultural subjects, the December 1842 meeting provided the local society with an opportunity to educate fellow South Carolina planters on the various experiments taking place on their plantations. Located in the middle of the St. John's Parish in Berkeley County, or Middle St. John's, the Black Oak Agricultural Society consisted of the most successful and prominent cotton and rice planters bordering the two branches of the Cooper River. Ravenel was their spokesperson. A planter of long-staple cotton, Ravenel represented one of many Middle St. John's planters whose ancestors had gained wealth from inland rice cultivation while the younger generation chose to pursue the lucrative cash crop.[1]

Ravenel's topics characterized the society's interest in the developing scientific agriculture most notably promoted by Virginia planter Edmund Ruffin. The report followed a systematic discussion of improving the agricultural conditions for growing the region's primary cash crop, Santee long-staple cotton. Ravenel first informed the audience of the "nature and quality" of St. John's soils, that the region had soil favorable to growing a

[1] "Agricultural Memoir," *Southern Agriculturalist* 3 (April 1843): 131; Minutes, March 1842, Black Oak Agricultural Society, South Carolina Historical Society (SCHS), Charleston, SC. St. John's Berkeley is not to be confused with St. John's Parish in Colleton County, established in 1730, on John's Island southwest of Charleston. In this chapter, any reference to St. John's refers to the parish Berkeley County.

variety of crops. The second topic, "applicability of Calcareous Manures to our Pine-land," reflected the scientific interests of improving soil fertility in long-staple cotton microenvironments. Planters' adaptation of new methods of fertilization (including compost, cotton seed, gypsum, and wood ash) also reflected how Ravenel and his peers were attempting to improve soil fertility in relation to cotton agriculture. Following his scientific analysis, Ravenel presented an overview of Santee long-staple cotton to qualify the crop's connection to the region. However, the final topic, titled "A succinct history of the former cultivation of Rice in our inland swamps, and the practicability of again resorting to the same culture, and its probable profits" strayed from the previous theme of soil improvement for long-staple cotton. Instead, his conclusion called for the revitalization of inland rice farming and tapped into reformers' calls for agricultural diversification and planter self-subsistence.[2]

Despite a call for reform efforts, Ravenel did not exhibit the qualities of book farming. Ravenel differed from Ruffin and the ideology of field rotation, believing that the process was too expensive and inefficient in managing his labor force. Instead, he tapped into the Middle St. John's planters' connection to the landscape and cultural history. Relying on his own and neighbors' scientific observations of marl as a fertilizer, plus drumming up cultural connections to past agricultural endeavors, Ravenel attempted to revive inland rice cultivation as a way to add to planter profits and avoid the unpredictable cotton-market swings. Stressing the connection of French Huguenots settling the area and their initial economic success from rice cultivation by the early eighteenth century, Ravenal lobbied for historical precedent to guide cotton planters to supplement their volatile plantation enterprise.

Ravenel's interpretation of how French Huguenots transformed a "foreign wilderness" into a profitable enterprise in Middle St. John's might have been applied to much of the agricultural history in the colonial Lowcountry. From the late seventeenth to the late eighteenth century, rice became a lucrative cash crop to Middle St. John's planters. Each generation of rice planters purchased an increasing number of slaves, who improved the inland rice infrastructure, as seen in water management and field networks. For example, the parish slave population rose from 180 to 1,439 people between 1705 and 1720. By 1790, the enslaved population grew to 9,900 people. To diversify their agricultural output,

[2] "Agricultural Memoir," 131.

Middle St. John's planters successfully augmented their low-lying crop by cultivating indigo on high ground by the 1740s. Following the Revolutionary War, political and market forces stifled the profitability of both crops. Indigo lost favor with South Carolina planters after Parliament terminated the bounty after American independence. Termination of the indigo bounty created a dramatic drop in domestic prices with the removal of inflated prices and merchants seeking out higher-quality dye in India. Also, the war left the abandoned inland rice embankments and canals in disrepair. The cost to repair and maintain these inland irrigation systems motivated the Middle St. John's planters to seek out other commercial options.[3]

For many of the local planters, the 1793 introduction of long-staple cotton in Middle St. John's provided answers to their agricultural problems. The crop was suited to the region's highland light sandy loam, where planters first experimented with the crop on abandoned indigo fields. Through a decade of trial and error, long-staple cotton became the primary cash crop between the Santee River and the upper boundary of tidal rice irrigation along the Cooper River at Moncks Corner, South Carolina. Long-staple cotton filled an agricultural niche above the residents' former inland rice fields. Despite Middle St. John's planters directing most of their enslaved labor to produce the crop, the regional long-staple variety could not compete with the finer quality Sea Island strain grown along the coastline. Lackluster seed selection, accidental hybridization with upland short-staple varieties, and the variation of soil content between Berkeley County and the sea islands were three reasons why planters could not produce the fineness of texture and length of Sea Island cotton. By 1818, Charleston merchants began distinguishing the

[3] Ibid; George D. Terry, "'Champaign Country:' A Social History of an Eighteenth Century Lowcountry Parish in South Carolina, St. Johns Berkeley County," (Ph.D. diss., University of South Carolina, 1981), 243, 254–255; "An Amount of the Number of the Inhabitants of the Parish of St. Johns, 1790," Census Records 1790–1930, SCDAH. For information on French Huguenots' settlement in South Carolina see: Daniel Ravenel, "Historical Sketch of Huguenot Congregations of South Carolina," *Transactions of the French Huguenot Society* 7 (1900): 7–74; Bertrand Van Ruymbeke, "Huguenots of Proprietary South Carolina: Patterns of Migration and Integration," in *Money, Trade, and Power: The Evolution of Colonial South Carolina's Plantation Society*, eds. Jack P. Greene, Rosemary Brana-Shute, and Randy J. Sparks (Columbia: University of South Carolina Press, 2001); Bertrand Van Ruymbeke, *From New Babylon to Eden: The Huguenots and Their Migration to Colonial South Carolina* (Columbia: University of South Carolina Press, 2006); Arthur H. Hirsch, *The Huguenots of South Carolina* (Durham: Duke University Press, 1928; reprint: Columbia: University of South Carolina Press, 1999).

long-staple cotton, called "Santee long," from the higher quality Sea Island variety with two price divisions.[4]

The inferior quality of the Middle St. John's long-staple cotton crop motivated planters to produce higher yields for a perceived higher accumulation of wealth. For example, the price difference between the two strains varied broadly from 2¢ to 20¢ per pound between 1816 and 1825. To prevent degeneration of their crop, planters in the Middle and Upper St. John's had to import annually new seed from the sea islands. Planters did not practice seed selection because they could not harvest first quality cotton seeds from their crop. The lack of seed selection eliminated the chance of a local variety taking hold in Berkeley County; local varieties were developed by planters who carefully experimented with plant strains better suited for specific environments. The lack of a local variety forced planters to import new Sea Island seeds every one to two years, which started the cycle all over again. To make up for the lack of profits obtained by a higher price per pound, Middle St. John's planters increased their output.[5] Samuel Dubose noted in 1858 that, "quantity and not quality was the aim in view; consequently, heavier yields were obtained from our lands."[6] By 1840, fluctuating cotton prices and nutrient-depleted soils hit these planters with diminished returns.[7]

A specific group of Lowcountry planters saw inland rice cultivation as an additional economic resource to counter the shifting cotton market in the late 1830s and 1840s. Situated upon a landscape of spring-fed wetlands and sandy-loam pinelands, a handful of Middle St. John's planters revitalized the agricultural mode of production that brought economic success to their ancestors. Inland rice farming provided notable wealth for the free residents living near the low-lying wetlands of Middle St. John's, called the Biggin Basin, during the eighteenth century.[8] Yet planters replaced the cash crop in favor of the emerging long-staple cotton boom

[4] Richard D. Porcher and Sarah Fick, *The Story of Sea Island Cotton* (Charleston: Wyrick Press, 2005), 108–116; Tamara Miner Haygood, *Henry William Ravenel, 1814–1887: South Carolina Scientist in the Civil War Era* (Tuscaloosa: University of Alabama Press, 1987), 27.

[5] Porcher and Fick, *Sea Island Cotton*, 108–116, table 13-2, 311.

[6] Ibid., 110.

[7] For lessons of cotton and soil depletion, see: Steven Stoll, *Larding the Lean Earth: Soil and Society in Nineteenth-Century America*, (New York: Hill and Wang, 2002), and Paul S. Sutter, "What Gullies Mean: Georgia's 'Little Grand Canyon' and Southern Environmental History," *Journal of Southern History* 76 (August 2010): 579–616.

[8] The Biggin Basin is now called the Pinopolis Basin, and is submerged under Lake Moultrie. For information on the history of the Santee-Cooper Navigation and Hydro-Electric Project and development of Lake Moultrie, see T. Robert Hart, "Santee-Cooper

in the early nineteenth century. The story of cotton agriculture in the region follows a similar path compared with the rest of the state. With rising market prices in the first two decades of the nineteenth century, planters aggressively planted cotton on the highland soil. However, declining cotton yields coupled with shifting market prices in the 1830s and 1840s caused regional planters to change their agricultural practices. Although these planters did not abandon cotton farming during the last two decades of the antebellum period, their production of inland rice explains how planters used one out of several crops to diversify their agricultural output as a means to maintain control in an evolving plantation enterprise.[9] According to agricultural historian William M. Mathew, diversification represented several strategies. Staple diversification allowed farmers to gain economic independence by growing a variety of crops to avoid being susceptible to market swings on one crop in particular. Planters – who could afford to do so – practiced area diversification when they grew crops on a variety of properties, either in the state or in the region. Also, sectorial diversification took place when landowners looked to supplement their income outside of farming, seen in mining or manufacturing.[10] Middle St. John's planters identified rice as one of several staples that could be grown to achieve agricultural diversity.

Biggin Basin planters represented the few South Carolina landowners who made a conscious effort to practice agricultural reform. As South Carolina cotton planters experienced diminishing returns, population emigration, and soil deterioration, dialogue appeared in agricultural societies and journals about the best methods to combat these problems. Characterized by Edmund Ruffin's call for improvement of soil fertility through a scientific approach to cultivation, planters used a combination of mixed husbandry and the application of calcareous marl to renourish impoverished soils created by decades of intensive farming. Despite reformers' efforts to spread the "gospel of marl," historians have

Landscape: Culture and Environment in the South Carolina Lowcountry," (Ph.D. diss., University of Alabama, 2004).

9 Porcher and Fick, *Sea Island Cotton*; Theodore Rosengarten, *Tombee: Portrait of a Cotton Planter* (New York: William Marrow, 1986); Alfred Glaze Smith, *Economic Readjustment of an Old Cotton State: South Carolina, 1820–1860* (Columbia: University of South Carolina Press, 1958); Lewis Cecil Gray, *History of Agriculture in the Southern United States to 1860*, 2 vols. (Washington, D.C.: Carnegie Institution of Washington, 1933; reprint Gloucester, MA: Peter Smith, 1958).

10 William M. Mathew, "The South Carolina Agricultural Survey of 1843," in *Agriculture, Geology, and Society in Antebellum South Carolina: The Private Diary of Edmund Ruffin, 1843*, ed. William M. Mathew (Athens: University of Georgia Press, 1992), 21.

documented South Carolina planters' unwillingness to adopt agricultural improvement in the antebellum period.[11] Ravenel and neighboring planters saw rice cultivation as one contribution to a larger ideology for agricultural reform. Rice became one of several crops grown by planters in order to move away from a mono-crop cotton culture, and the basin became one of few areas in the state to initiate reform. Southern historian William Scarborough argues, "only large planters applied scientific practices who had sufficient leisure, education, and familiarity with the literature to make science the true handmaid of agriculture." Scarborough categorized these large planters into three divisions of scientific familiarity, with "planter-scientists" representing the upper echelon of this class. Ravenel fell into this category with an education in the sciences and bountiful income derived from inheritance into the planter elite.[12]

Black Oak Agricultural Society members exchanged ideas about how to deal with the dilemma of declining cotton productivity. Inspired by Justus von Liebig, Edmund Ruffin, and other "agricultural Chemists," Ravenel's address reveals how agricultural reformers attempted to use the emerging nineteenth century soil sciences as a way of solving the strained relations between enslaved agriculture and society. Improving farming conditions, according to Ravenel, encouraged more South Carolinians to

[11] For scholarship documenting South Carolina planters and the failure of agricultural reform, see: William M. Mathew, *Edmund Ruffin and the Crisis of Slavery: The Failure of Agricultural Reform* (Athens: University of Georgia Press, 1988), specifically 206–208; Stoll, *Larding the Lean Earth*, 155–160; Clement Eaton, *The Mind of the Old South* (Baton Rouge: Louisiana State University Press, 1964), 137, 156–157; Gray, *Agriculture in the Southern United States*, vol. II, 788–789; Smith, *Economic Readjustment*, 69–111; Charles G. Steffen, "In Search of the Good Overseer: The Failure of the Agricultural Reform Movement in Lowcountry South Carolina, 1821–1834," *Journal of Southern History* 63 (November 1997): 753–802; Steven G. Collins, "System, Organization, and Agricultural Reform in the Antebellum South, 1840–1860," *Agricultural History* 75 (Winter 2001): 1–27; Drew Gilpin Faust, "Rhetoric and Ritual and Agriculture in Antebellum South Carolina," *Journal of Southern History* 45 (November 1979): 541–568. For analysis of Southern agricultural reform, see: John D. Majewski, *Modernizing a Slave Economy: The Economic Vision of the Confederate Nation* (Chapel Hill: University of North Carolina Press, 2009), 56–61, 73–74, 227–228. William Scarborough argues planter elites embraced the agricultural reform movement and developing scientific theories. However, smaller planters and yeoman farmers failed to implement agricultural reform; in Scarborough, "Science on the Plantation," in *Science and Medicine in the Old South*, eds. Ronald L. Numbers and Todd L. Savitt (Baton Rouge: Louisiana State Press, 1989), 95–106. Lacy Ford counters this argument, saying that upcountry Yeomen maintained self sufficiency through a diverse market of foodstuffs; in Ford, "Self-Sufficiency, Cotton, and Economic Development in the South Carolina Upcountry, 1800–1860," *Journal of Economic History* 45 (June 1985): 261–267.

[12] Scarborough, "Science on the Plantation," 87.

remain in the state instead of migrating westward to seek out cheap and fertile land. Ravenel constructed an argument for why planters should practice soil fertilization. He understood the ecology of Middle St. John's by associating soil content with plant growth. Although soil sciences were still in their infancy, Ravenel argued that lighter, sandier soils supported pines and oaks "in their virgin state" and provided ideal conditions for cotton, while the heavier moist soils supporting cypress and gums were better for rice. While people identified suitable agricultural regions by associating particular soils with the forests that grew on them in the colonial era, Ravenel actually attempted to explain how soil "ingredients" (carbon, oxygen, hydrogen, and nitrogen) provided the nutrients for different plants to grow. For Ravenel's audience, the removal of nutrients from the soil also had lasting consequences. The heart of Ravenel's paper documented how intensive agriculture had caused the decline of regional soil fertility and how various fertilizers, or "manures," could rejuvenate specific soil associations.[13]

Along with soil rejuvenation, Ravenel also saw agricultural diversity as a way for cotton farmers to overcome economic depressions. Ravenel explained that this region's soil was quite capable of successfully producing crops other than cotton and corn. He stressed that southern planters should diversify their agriculture by suggesting a variety of crops historically grown in the state. Indigo, rice, silk, tobacco, wheat, and castor beans could supplement planters' income when faced with depressions in the cotton market. Ravenel concluded his lecture with an emphasis on inland rice cultivation and its historical connection to Middle St. John's. By the 1840s, rice prices were relatively stable compared to the oscillating cotton market. He believed that fellow residents could effectively restore the abandoned inland rice fields. Planters could efficiently transport the rice to Charleston by waterways, specifically the Santee Canal, already in place from the internal improvement boom of the early nineteenth century. Envisioning rice production as a solution to the Biggin Basin's agricultural woes, Ravenel exclaimed that former lands would "draw again to her bosom her alienated sons, and the gracious promise of old, will come to gladden her fields with joy." To these planters living in uncertain financial times, rice served as a symbol for future economic stability.[14]

[13] "Agricultural Memoir," 131; Stoll, *Larding the Lean Earth*, 150–155.

[14] "Agricultural Memoir," quote, 131; Haygood, *Henry William Ravenel*, 29–30; Rosengarten, "Southern Agriculturalist," 281–284.

At the heart of this story is the role that irrigation and drainage canals played in Middle St. John's history. Five generations of planters developed and maintained this infrastructure to control the tremendous amount of water flow distinctive to the basin. A high concentration of springs dotted the region, with streams flowing through the Biggin Basin and converging into centrally located swamps. The vast acreage of low-lying wetlands prompted residents to devise intricate and sprawling canal systems to effectively cultivate and drain the region. Planters first focused their attention at the turn of the eighteenth century on altering the landscape to cultivate rice. With the rise of the cotton economy and resulting movement to improve the state transportation infrastructure, enslaved labor changed the Biggin Basin hydrology once again. The Santee Canal cut through the middle of the basin and contributed to the rise of the region's reliance on mono-crop agriculture by providing an efficient way to transport cotton to Charleston. While residents of Middle St. John's committed to cultivating long-staple cotton, the surrounding drains and canals provided constant reminders of the region's inland rice history. As planters began discussing possibilities for agricultural diversification, these former irrigation networks made the possibility of reviving inland rice production appear possible.

Middle St. John's is one of the ten original parishes created by South Carolina's General Assembly between 1706 and 1708 and its boundary stretched fifty miles from the East Branch of the Cooper River to the Santee River.[15] By the antebellum period, residents began designating Lower, Middle, and Upper St. John's Parish.[16] The Biggin Basin formed the topographical heart of Middle St. John's. Named after Biggin Hill in Kent, England by Landgrave John Colleton, the basin consisted of four tributaries converging into Biggin Swamp. The swamp formed the headwaters of the Western Branch of the Cooper River. The unique geological and hydrological formations of Middle St. John's attracted plantation development by the early eighteenth century. More than 40 million years ago shells and calcium carbonate lime sediments were deposited over the

[15] "Account of an Agricultural Excursion into St. John's, Berkley," *Southern Cabinet of Agriculture, Horticulture, Rural and Domestic Economy* 1 (January 1840): 13; Terry, "'Champaign Country," 16–17.

[16] Harriett Kershaw Leiding, *Historic Houses of South Carolina* (Philadelphia: Lipencott & Co., 1921), 132.

area. These deposits are now known as the Santee Limestone. It was these deposits, buried in some cases three to five feet under inland swamps, which attracted Edmund Ruffin to study this region. The 60,000-acre basin lay between the Dorchester and Summerville scarps, two of many such scarps in the Lowcountry, which served as approximate boundary lines of Middle St. John's.[17] The Dorchester Scarp formed the northern and western borders of the Biggin Basin. It is the seaward limit of an early Pleistocene barrier island system. The Summerville Scarp is the seaward limit of the next younger barrier island system. This barrier island system, that underlies modern-day Bonneau Beach, Pinopolis, and Summerville, formed the eastern border. Over the Santee Limestone deposits, or marl, a clay loam formed from backbarrier deposits on the landward side of the Summerville Scarp's associated barrier islands. This loam provided ideal water impermeability that aspiring rice planters would seek out in the eighteenth century.[18]

Residents' colonial development of Biggin Basin rice cultivation and resulting irrigation systems reflected the close association of the plantation economy and the local topography. Although New England Anabaptists, English, Barbadians, Irish, and enslaved Africans settled in the region, French Huguenots made up the majority of the local population. By the eve of the Revolutionary War, fifteen Biggin Basin plantations were under the direction of the Ravenel, Porcher, Mazyck, and Moultrie families. The high concentration of limestone springs created a uniquely conducive environment for inland rice cultivation in the low-lying areas. Part of the Floridian aquifer system's northeastern boundary, the limestone springs stretched across South Carolina's southwestern Lower Coastal Plain. This formation appears in upper Berkeley County, next to the Santee River, and extends toward the Savannah River and into Georgia. The Santee Limestone formation was part of a series of "calcareous deposits" that would become an important ingredient for

17 W.R. Doar, III, and Ralph H. Willoughby, *Revision of the Pleistocene Dorchester and Summerville Scarps, the Inland Limits of the Penholoway Terrace, Central South Carolina* (Columbia: South Carolina Department of Natural Resources, Geological Survey, 2006); Donald J. Colquhoun and Mark L. Brooks, "New Evidence from the Southeastern U.S. for Eustatic Components in the Late Holocene Sea Levels," *Geoarchaeology* 1 (April 1986): 284; Suzanne Cameron Linder, *Anglican Churches in Colonial South Carolina: Their History and Architecture* (Charleston: Wyrick & Co., 2000), 60.

18 W.R. Doar, III, "Advice," email, 2018; Doar and Willoughby, *Pleistocene Dorchester and Summerville Scarps*; Colquhoun and Brooks, "New Evidence from the Southeastern U.S.," 284.

agricultural reformers in the antebellum era.[19] Early geologists named this formation after the exposures along the eroded Santee River beds in upper Berkeley County.[20] During the Oligocene and Miocene, the Santee Limestone was exposed to "subsurface erosion" that formed "caverns and conduits" through the rock. Clay and siliceous phosphates, that would become instrumental in rice field water retention, covered the limestone formations during the Miocene and served as a cap to the porous layer. Downdips, or sinks, in the Santee Limestone occurred from the upper sections of limestone and were prevalent throughout the Biggin Basin.[21]

Unlike inland planters who relied on rainfall collected upstream in impounded reservoirs, Biggin Basin planters could tap into artesian wells and harness ample water supplies.[22] Woodboo had the most extensive spring fed irrigation in Middle St. John's. Drawing from two major springs on the property and impounding spring fed water from a neighboring plantation, the Mazyck family managed this constant water supply to form reservoirs and canal systems directing water through the fields and into the Biggin Creek and the Santee Canal. On the eastern perimeter of Biggin Swamp, just below the Woodboo settlement highlands overlooking the rice fields, lay "Big Spring." In a scientific paper describing the limestone springs of St. John's, Dr. Edmund Ravenel of neighboring Hog Swamp Plantation stated that the Woodboo's Big Spring was "the most remarkable spring in this locality." Big Spring had a depth of over thirty feet with another spring tributary feeding into the watercourse two

[19] Christian Senf, "General Plan of the Canal and Its Environs between the Santee and Cooper Rivers in the State of South Carolina," ca.1800, Maps and Plats File, SCHS; Walter R. Aucott and Gary K. Speiran, "Ground-Water Flow in the Coastal Plain Aquifers of South Carolina," *Ground Water* 23 (November/December, 1985): 738. The Floridian aquifer system covers approximately 100,000 square miles in South Carolina, Georgia, Alabama, Mississippi, and Florida.

[20] C. Wythe Cooke, *Geology of the Coastal Plain of South Carolina*, Bulletin 867 (Washington, D.C.: United States Geological Survey, 1936), 75.

[21] George E. Siple, "Some Geologic and Hydrologic Factors Affecting Limestone Terraces of Tertiary Age in South Carolina," *Southeastern Geology* 2 (August 1960): 9; Walter R. Aucott, "The Predevelopment Ground-Water Flow System & Hydraulic Characteristics of the Coastal Plain Aquifers of South Carolina," USGS Water Investigations Report B6–4347 (Washington, D.C.: GPO, 1988), 9.

[22] Edmund Ravenel, "The Limestone Springs of St. John's Berkeley," *Proceedings of the Elliott Society* (October 1860): 28–31; F.S. Holms, "Notes on the Geology of Charleston, SC," *American Journal of Science and Arts* 7 (March 1849): 199–201; Judith Carney and Richard Porcher, "Geographies of the Past: Rice, Slaves, and Technological Transfer in South Carolina," *Southeastern Geographer* 33 (November 1993): 138–139.

hundred yards down stream with a depth of forty feet. Although Big Spring was only five feet wide, its water flow was strong enough to produce currents and eddies. Ravenel also reported that Woodboo's second spring, "[had] never been known to fail." The Mazycks created a reservoir from this stream and enabled them to generate a steady flow of water down to rice fields below.[23]

The Floridian aquifer's water flow enabled the Mazycks to cultivate rice with relative confidence by avoiding drought. By 1806, Woodboo had eighty-four acres of rice fields under spring fed irrigation. The clear springs flowing through the property inspired Pierre de St. Julien de Malacre to call the surrounding land "Belle Fontaine," or "beautiful fountain." The Mazyck rice fields were a simple rectangular formation sandwiched between high ground on the northeast and Biggin Creek and Santee Canal on the southwest. Twenty field divisions managed water flow more systematically to flood and drain water precisely in a little over a four-acre plot.[24]

Big Spring's location and consistent discharge served a critical role in Woodboo's successful irrigation. This water flow provided enough water to disperse through the downward sloping fields, moving away from the settlement's high land and toward the Biggin Creek basin. Halfway down the northeast side of the rice field system impounded water entered from the Woodboo reservoir. This water could flood about half of the field system in a similar manner by dispersing water from the northern point bisecting the fields, and flowing through the check canals to the smaller field divisions. The springs flowed so well that late nineteenth century College of Charleston "History & Belles Lettres" professor, and Somerton Plantation owner, Frederick Adolphus Porcher noted that this irrigation system could cover forty acres of rice in one night.[25]

At the confluence of the swamp's four tributaries lay eleven plantations interconnected through irrigation and drainage systems. The Ophir, Moultrie, and Pooshee Canals served as the central drainage arteries of Biggin Swamp. Individual plantation owners constructed these canal

[23] Ravenel, "Limestone Springs," 28–29, Holms "Notes on the Geology," 187; "Plat of Woodboo," 1806, Maps and Plats File, SCHS. For more information on Woodboo Springs, see: "Glimpses at the Country of the Olden Time," *Russell's Magazine* 2 (October 1857): 64. The Mazyck family subdivided Woodboo plantation into Woodboo and Fair Springs plantations.

[24] "Plat of Woodboo," SCHS.

[25] Ibid.; Frederick A. Porcher, "Upper Beat of South Carolina: A Memoir," *Transactions of the Huguenot Society of South Carolina* 13 (1906): 46.

systems through their property, linking them together to form an interconnected irrigation and drainage system. Ophir Canal ran through the middle of Ferguson Swamp, intersecting with at least four plantations. The canal helped irrigate 450 acres in Ophir Plantation and powered the plantation rice mill.[26] Ophir Canal intersected with Moultrie Canal and passed through Biggin Swamp. Moultrie Canal began three plantations to the north of its confluence with Ophir, as the two continued through a series of plantations toward Moncks Corner. On the east tributary of lower Biggin Swamp lay Pooshee Canal, a spring fed system that connected to the Moultrie Canal.[27] As cotton agriculture took hold in the region by the early nineteenth century, the Biggin Swamp canal system became a nuisance, providing a conduit for freshets that breached former rice embankments and destroyed bisecting bridges and causeways.[28] By the early 1840s, Ravenel's interest in rice redefined this arterial waterway as an important component providing irrigation to neighboring fields.

Despite the ample water available to irrigate inland rice fields, Biggin Basin planters had difficulty transporting their crop to Charleston because the navigability of the Cooper River stopped at the Summerville Scarp, located approximately thirty-five miles upriver at Stoney Landing. To improve their condition, Middle St. John's residents sought internal improvements for navigation soon after their settlement to the area.[29] Construction on the Biggin Swamp canal system began in 1702 after local residents successfully lobbied the Commons House of Assembly the previous year for internal improvements. Motivated by extending the navigability of the West Branch of the Cooper River above Fairlawn Barony, Biggin Swamp residents pushed to have the colonial government fund legislation "making and mending High Ways and paths and for cutting of Creek & Water Courses" the following year. This legislation caused some plantation owners, who feared that an altered watercourse would hinder their ability to irrigate rice, to protest. Proponents argued that the Biggin Canal provided beneficial navigation to residents and did

[26] "Plat of Ophir Plantation," 1931, J.P. Gaillard Plat Collection, CCRMC.

[27] Chicora quadrangle, 1921; US Geological Survey, South Carolina [map], 1:62,500, 15 Minute Series (Washington, D.C.: US Department of the Interior, USGS, 1921); Senf, "General Plan of the Canal and Its Environs," SCHS.

[28] Samuel Porcher to Isaac Dwight, 22 Sept 1837, 16 June 1840, Isaac M. Dwight Family Papers, SCHS.

[29] Terry, "Champaign Country," 52–53.

not hamper established rice fields.[30] In 1726, the General Assembly granted local inhabitants an act to extend navigability above Pooshee Plantation. Compared to improvements on individual plantations, the colony saw these projects benefiting the community as a whole. The legislature assigned commissioners to design and supervise the cutting and clearing of the creeks, and recruited "all and every [one of] the male inhabitants, from sixteen years old to sixty years" within the region to help dig the canal.[31]

In the early nineteenth century, the Santee Canal further altered the natural watercourses forming Biggin Swamp and symbolized how transportation demands of mid-state cotton planters overtook the infrastructure of inland rice planters. Serving as a waterway connecting the Santee River to the Cooper, the canal provided transportation to and from the South Carolina midlands, ultimately linking the three-year-old capital Columbia with Charleston, the largest southern seaport during the Early Republic. Besides the limestone springs, the Santee Canal was the second feature that dramatically set apart irrigation practices of this region as compared to their counterparts down river. The canal was twenty-two miles long, thirty-two feet wide and up to eight feet deep, connecting the upper Santee River to the Cooper River. An example of the new country's quest for internal improvements, the Santee Canal connected Columbia to Charleston through a continuous watercourse. Its construction occurred during the first canal boom between the Revolutionary War and the War of 1812, when chartered companies privately funded still-water canals to serve as toll-based connections between strategic rivers.[32]

While Biggin Basin planters originally opposed the canal's construction because the waterway would disrupt farming and alter plantation boundaries, the canal's transportation capabilities and additional water for irrigation led many residents to embrace the environmental alteration as an engineering marvel during the early antebellum era.[33] The canal

[30] Ibid., 179–180; *Journals of the Commons House of Assembly of South Carolina for 1702*, ed. A.S. Salley (Columbia: Printed for the Historical Commission of South Carolina by the State Company, 1932), 26.

[31] "An Act to Empower Several Commissioners," Act 442, 15 September 1721; "An Act for Cutting and Clearing a Creek, Commonly Called Biggon Creek," Act 417, 11 March 1726; in *Statutes at Large for South Carolina*, vol. 9, ed. David J. McCord (Columbia: A.S. Johnson, 1841).

[32] Robert J. Kapsch, *Historic Canals and Waterways of South Carolina* (Columbia: University of South Carolina Press, 2010), 6–7.

[33] Mabel L. Webber, ed., "Col. Senf's Account of the Santee Canal," *South Carolina Historical and Genealogical Magazine* 28 (January 1927): 10; Samuel Gaillard Stony,

followed the Biggin Swamp floodplain in the lower sections, yet the upper section traversed higher ground separating the Cooper and Santee watersheds. To counter the thirty-four foot rise over the barrier complex leading up to the Dorchester Scarp, engineers under the direction of State Engineer Colonel Christian Senf created ten locks consisting of brick enclosures sixty-feet long with cypress gates. Slaves controlled water from adjoining reservoirs or slues to generate or release water pressure, to raise or lower barges from the Santee floodplains to the headwaters of the Cooper. The Santee Canal served as a major artery through the central South Carolina Lowcountry between 1800 and 1853; however, the emerging railroad networks, the canal's limited payload, and repeated problems during droughts forced the Santee Canal Company to abandon operations in 1853.[34]

The Santee Canal development provided additional income to regional planters while Middle St. John's transitioned from a rice economy to a cotton economy. Construction took place between 1793 and 1800, which was the decade when local residents began experimenting with cotton production. With the rice production all but ceased, planters generated income hiring out their slaves to the Santee Canal Company. Six months into the project, Charles Cotesworth Pinckney wrote, "we have eight hundred Negroes at work on the canal, yet it goes on very slowly."[35] In 1793, the company paid slaveholders £15–£16 per slave annually. By 1800, with planters moving their slaves back to the plantations to grow cotton, the Santee Canal Company raised the payment to £20–£24 a slave. With most planters in Middle St. John's devoting at least some of their enslaved labor to work on the project, slaves' wages provided planters subsistence while they figured out how to cultivate the highland cash crop. Rene Ravenel of Pooshee noted in his diary on January 1, 1796 that, "my Negroes return'd home from the Santee Canal after working three years," which supplemented his income before cotton became the primary cash crop. The enslaved labor population working on the canal

ed., "Memoirs of Frederick Augustus Porcher," *South Carolina Historical and Genealogical Magazine* 44 (April 1943): 70–71.

[34] For further history of the Santee Canal, see: Kapsch, *Historic Canals*; F.A. Porcher, "The History of the Santee Canal," (Charleston: South Carolina Historical Society, 1875; reprint, 1903); Henry S. Tanner, *A Brief Description of the Canals and Railroads of the United States* (Philadelphia: Published by the Author, 1834); Hart, "Santee-Cooper Landscape," 21–24.

[35] Charles Cotesworth Pinckney to "my dear brother," 28 November 1793, Pinckney Family Papers, SCHS.

grew from 150 to 1,000 people in 1793. By 1796, the labor force was down to 700 people. Yet from 1793 to 1800, planters reaped approximately £220,000 by hiring out their slaves to the Santee Canal Company.[36]

The canal's controversial plan became a divisive marker between cotton and rice planters. The Santee Canal Company consisted of an elite political base pursuing South Carolina internal improvements after the Revolutionary War, and it attempted to capitalize on the growing cotton market. General William Moultrie, who introduced long-staple cotton to Middle St. John's in 1793, served as the company's president. His peers and former Revolutionary War officers composed the board of directors. Conveniently, the canal passed by Moultrie's two principal plantations without the route impinging upon his rice or cotton fields. Moultrie thus had the luxury of transporting agricultural commodities to Charleston without the canal destroying any of his agricultural land.[37]

Neighboring rice planters did not share in Moultrie's fortune. The canal repeatedly bisected Biggin Swamp and its tributaries, affecting almost every plantation bordering the waterway. For those planters, the canal cut their property lines or rice fields. As a result, planters sought to readjust boundary lines around the new canal and adapt to alterations in irrigation and drainage. The canal severed parcels of inland rice fields from central plantation tracts. Planters affected by this dissection of property attempted to trade acreage with one another. Using the canal as a new boundary line, planters attempted to annex rice fields from neighboring plantations that now lay on the their side of the division. In some cases, planters cordially exchanged property. Daniel Ravenel of Wantoot, for example, exchanged sixteen acres for an equal number with Stephen Mazyck of Woodboo. In other cases, planters scrambled to make up lost acreage from the canal bisecting their rice fields.[38]

As the Santee Canal route altered property boundaries, some planters chose to sell large tracts severed from their original plantation. The Mazyck family sold 198 acres of Woodboo and Fair Spring, severed from the plantations by the Santee Canal, to William Cain of adjoining Somerton Plantation.[39] Some planters chose to maintain valuable rice lands as an annex to their central property. Daniel Ravenel of Wantoot and Rene Ravenel of Pooshee continued to hold properties after the canal

[36] Porcher, "Santee Canal," 4–5; 1 January 1796, "Diary of Henry and Rene Ravenel," Ravenel Family Papers, SCHS.

[37] Kapsch, *Historic Canals*, 24–27.

[38] "Plat of Woodboo," SCHS.

[39] "Ibid.

severed hundreds of acres off their western boundaries.[40] Not only did the canal run through rice fields, but the canal's positioning in specific circumstances rendered some land useless without proper irrigation or drainage. For example, Daniel Ravenel lost two and a half acres of rice fields from the canal construction. Downstream neighbor Stephen Mazyck faired worse than Ravenel, losing an additional ten acres to the canal's path.[41] These nuisances caused by the newly placed canal, according to Senf, were a small price to pay for "safe and convenient Navigation to market at any time." Senf further boasted, in a letter to the Santee Canal Company President William Moultrie, "in Justice such a planter should rather make some Compensation to the Company, than in place to receive any, who procure to him such great Advantages."[42]

The Santee Canal intensified Biggin Swamp freshets, making fledgling inland rice planters more susceptible to the natural disasters. Because the canal's large embankments compounded floodwaters down stream, similar to twentieth century levees inducing floods down river on the Mississippi River, planters consistently battled freshets after the canal's completion in 1800. Rene Ravenel of Pooshee recorded in 1800 that a heavy rain storm created a "higher Fresh in the Swamp than has been since last October," washing away bridges, and "the water run'd over the Bank of the Santee Canal in several places and made a breach in the Bank Ten Feet." He added, "the Santee Canal Bank broke in three places in Wantoot, and carry'd away about 25 or 30 Feet of the Bank in each place ... This is the highest fresh that had been for several years."[43] Rene states in his diary that the Santee Canal banks breached five times between 1800 and 1820, and his rice fields were inundated twice with freshets.[44]

40 Senf, "General Plan of the Canal and Its Environs," SCHS; Stony, "Memoirs of Frederick Augustus Porcher," 70; "Plan of Wantoot, Somerton, and Hog Swamp Plantations," 1 April 1795, Plat Book A, CCRMC; "Map of Pooshee," 1920, Berkeley County Register of Mesne Conveyance (BCRMC), Monks Corner, SC.

41 Webber, "Senf's Account," 17–18; "Robert McKelvy vs. Santee Canal Co.," 13 March 1800, Judgment Roll, South Carolina Department of Archives and History (SCDAH), Columbia, SC; "Theodore Gourdin vs. Santee Canal Co.," 11 June 1803, Judgment Roll, SCDAH.

42 Webber, "Senf's Account," 117.

43 24 June 1800, "Diary of Henry and Rene Ravenel, 1785–1851," Ravenel Family Papers, SCHS.

44 Ibid., 24 June 1800, 19–20 May 1806, 7–8 March 1807, 8–9 March 1810, 27 August 1813, 19 May 1817, 10 September 1820.

During a September 1837 storm, Samuel Porcher recorded that the Santee Canal's terminus was eight feet underwater at Biggin Bridge.[45]

Despite these human-induced disasters, inland rice planters benefited, in some cases, from the Santee Canal's irrigation systems.[46] Planters negotiated with canal engineers to keep central slues and irrigation-ways open to feed water from reservoirs and springs to the severed rice fields. As the canal dropped thirty-four feet from high sandy pinelands into the Biggin Swamp floodplain, the canal's route significantly affected planters below Pooshee. Daniel Ravenel worked with the canal engineers to install two floodgates that drained water out of the Wantoot rice fields and into the canal. Engineers designed four overflow canals that siphoned water away from the central Santee Canal during freshets, relieving any pressure from embankments adjoining the rice fields. Ravenel benefited from a reservoir created by the Santee Canal Company that captured Pooshee springs and sent the water to Wantoot's upper rice fields.[47]

Biggin Swamp planters had to manage water around the new channel creatively. For example, the Santee Canal severed the limestone spring-fed water flow to the Wantoot rice fields. To solve this problem, slaves built a brick aqueduct to carry the spring-fed water under the Santee Canal and into the Wantoot rice fields. The floodgates positioned along the canal banks enabled Ravenel's trunk minders to pull off water from the canal, where it merged with older Wantoot irrigation canals and into the rice fields. At the same time, two of the overflows emptied excess water into the "old side Drain" which was a central irrigation artery pulling water from the limestone springs and Biggin Creek.[48] A similar scenario existed for the two preceding plantations, Woodboo and Oakfield, which tapped into the canal's aqueduct to pull water through floodgates and overflows. Senf's report specifically addressed how the floodgates provided Woodboo's rice fields with water "in dry seasons." Mazyck's neighbor, Charles Johnson, had floodgates constructed "that the Proprietor of this land, who was supplied with Water from Mazycke's [sic], and other Springs, in dry seasons may now have the Use of the Surplus Water from the Canal." Senf stated that the canal brought more consistent water flow to the Oakfield floodgate because Johnson's rice fields were on a higher

45 Samuel Porcher to Isaac M. Dwight, 22 September 1837, Isaac M. Dwight Papers, SCHS.

46 Stony, "Memoirs of Frederick Augustus Porcher," 71; Hart "Santee-Cooper Landscape," 23–24.

47 Webber, "Senf's Account," 17–18.

48 Ibid., 17; Ravenel, "Limestone Springs," 29.

elevation that the canal could reach, unlike the lower flowing Biggin Creek.[49]

Just as planters took advantage of increased irrigation, the Santee Canal improved drainage of neighboring rice fields. For example, Ravenel drained two hundred acres of Wantoot rice fields by connecting his ditches to the canal. Eight miles of private canals converged with the Santee Canal downstream from Ravenel on Woodboo Plantation. The confluence of the canals magnified the freshets during substantial rains. To solve this problem, canal engineers specifically reinforced the western canal embankment to battle the "Debordement of the Creek," making the Woodboo side "four feet higher and five feet thicker" than the east bank. As a trade-off for the Santee Canal bisecting the Woodboo fields, the canal engineers constructed stronger embankments on Mazyck's property. Prior to construction, as Senf described, the Woodboo rice fields were "under very weak banks."[50]

Besides strengthening embankments, Santee Canal engineers improved floodgates that allowed planters to discharge water into the canal while adding to the canal's water volume for navigability. Senf's strategy paid off by tapping into the spring-fed channels, as the watercourse below Black Oak Lock consistently offered navigation. The "great drought of 1818," which actually consisted of annual droughts between 1817 and 1819, dried up many of the retaining reservoirs between White Oak and Black Oak, making the canal impassable for fifteen miles in traversing the Dorchester Scarp. Rene Ravenel of Pooshee described the drought in June 1819, where "no rain ... since November last, the big swamp so dry that but one creek had water, and that scarcely run." Despite the drought leaving the upper canal dry, water released from Pooshee, Wantoot, and Woodboo floodgates enabled the canal's southern section to remain open during this three-year drought.[51]

Biggin Basin planters began to abandon rice cultivation in exchange for long-staple cotton by the time William Buford made his inaugural voyage through the Santee Canal in May 1800. Beginning with General William Moultrie's first crop in 1793 and continuing until emancipation, cotton

[49] Webber, "Senf's Account," 18, 117. [50] Ibid., 18, 117; quote: 18.

[51] Ravenel, "Limestone Springs," 28, 29; Porcher, "History of the Santee Canal," 9, 11; Hart, "Santee-Cooper Landscape," 23; Holms, "Notes on the Geology," 187; 16 June 1819, "Diary of Henry and Rene Ravenel," Ravenel Family Papers, SCHS.

production moved Middle St. John's planters away from the laborious process of manipulating water for rice irrigation. Instead, these planters and their slaves had to readjust their understanding of agriculture from low wetlands to highland pine-barrens. The highland microenvironment presented new challenges with soil. Unbeknownst to the pioneering cotton planters, the sandy ridgelines lost nutrients faster than the low spring-fed watersheds. So while post-Revolution planters thought the new cash crop would relieve their economic problems, they in turn created new challenges in maintaining an intensive mono-crop environment on the high land.[52]

Long-staple cotton took hold in Middle St. John's for two reasons. The first was the state's loss of the indigo market share by the final decade of the eighteenth century. Parliament's repeal of a lucrative colonial bounty, British merchants' criticism of the poor quality in domestic dye, competition from French and Spanish Caribbean plantations, and a "flooding of the market by inexpensive, high-quality indigo from India" ultimately collapsed domestic production.[53] Either by observation or coincidence, Lowcountry planters recognized similarities between the indigo and cotton environments, as both crops thrived in a permeable sandy and nutrient-rich soil. Long-staple, or Sea Island, cotton was a fibrous variety which earned higher prices from English cotton factors. Planters first started cultivating this strain on the Georgia Sea Islands by 1789. Long-staple cotton worked its way up the South Carolina barrier island chain before William Moultrie attempted, and failed, to grow the strain on his inland plantation in 1793. Peter Gaillard at the Rocks Plantation on the upper Santee produced the first regional crop in 1796, and by 1800 several planters were either juggling inland rice cultivation with cotton production or abandoning the former cash crop altogether in favor of the latter.[54]

Middle St. John's planters' willingness to abandon inland rice cultivation in favor of long-staple cotton speaks to the changes in the plantation economy at the turn of the nineteenth century. Joyce Chaplin writes how

[52] Kapsch, *Historic Canals*, 47; Porcher and Fick, *Sea Island Cotton*, 108–116; Hart, "Santee-Cooper Landscape," 24–28; Abraham Blanding, "Report to the Superintendent of Public Works," 378, Committee Reports, SCDAH.

[53] Walter Edgar, *South Carolina: A History* (Columbia: University of South Carolina Press, 1998), 266. For South Carolina indigo production, see: Andrea Feeser, *Red, White, and Black Make Blue: Indigo in the Fabric of Colonial South Carolina Life* (Athens: University of Georgia Press, 2013).

[54] Porcher and Fick, *Sea Island Cotton*, 110–111; Terry, "Champaign Country," 356–360.

planters continued to plant rice after the Revolutionary War, not willing to abandon the crop for new forms of agricultural experimentation. But for Biggin Basin planters, the opportunities to capitalize from long-staple cotton outweighed their ability to continue rice cultivation. With the collapse of the indigo market after the war, regional planters were left with rice as a single cash crop. Declining inland rice yields in Middle St. John's coupled with inconsistent price fluctuations and devastating freshets served as motivation to begin a new agricultural venture. The Biggin Basin planters did not live close to tidal influenced floodplains on the Cooper River. Unlike Lower St. John's planters, who had the luxury of converting their rice fields into tidal irrigated systems, inland planters were limited in their agricultural choices. They could continue the older method of planting rice with declining output or they could attempt to grow a new cash crop with potentially high returns.[55] Even in some cases where older generations continued to manage inland rice, the younger generation set the practice aside to focus exclusively on long-staple cotton. Rene Ravenel, for instance, grew rice and cotton on his Pooshee and Indianfield plantations up until his death in 1822, yet Rene's son Henry made no mention of rice cultivation in his 1826 records or his plantation journals during the 1830s.[56]

With suitable land and enslaved labor already in hand, planters nestled around the Biggin Basin who switched to cotton quickly saw high rates of return. But as soon as these cotton planters saw success in their endeavors, they experienced the market's unpredictability. For example, during the first three decades of the nineteenth century Santee long-staple cotton reached a peak of 60¢ per pound in 1818 and a low of 19¢ per pound in 1823.[57] By 1846, twenty-seven plantations in Biggin Basin devoted an average of 135 acres to cotton with Pooshee and Somerset growing 300 acres each. As the editor of the *Southern Cultivator* observed of the St. John's scene, "it is seldom that a field, selected for cotton in this

[55] Joyce E. Chaplin, *An Anxious Pursuit: Agricultural Innovation and Modernity in the Lower South, 1730–1815* (Chapel Hill: University of North Carolina Press, 1993), 226.

[56] J. L. Smith, "Analysis of the Cotton Lands on the Head Waters of the Cooper River," in *Report on the Geology of South Carolina*, M. Toumey (Columbia: A.S. Johnston, 1848), xliii–xlvii; "Account of an Agricultural Excursion," 78; "Diary of Henry and Rene Ravenel," SCHS; "Henry Ravenel Crop Book," Ravenel Family Papers, SCHS, "Henry Ravenel Receipts, 1823–1854," Ravenel Family Papers, SCHS; Porcher and Fick, *Sea Island Cotton*, 116. Quotes: "Agricultural Memoir," 131.

[57] Porcher and Fick, *Sea Island Cotton*, 311; Gray, *History of Agriculture*, vol. II, 1031; long-staple cotton prices even higher in Dr. Henry Ravenel's Pooshee plantation journal, Ravenel Family Papers, SCHS.

Parish, is cultivated in any other crop." Planters devoted increasing labor and acreage to growing cotton. Even during depressions in the cotton market, planters continued to increase their crop output according to South Carolina historian Alfred Glaze Smith, Jr. "with the thought ever in their mind ... that 'next year' would bring long awaited success."[58]

During the second quarter of the nineteenth century, Middle St. John's cotton planters found themselves in the same dilemma as the rest of their southern contemporaries. Dramatic fluctuations in the cotton prices combined with declining cotton yields created a sense of urgency for Middle St. John's planters. For a vast majority of the regional planters, their desire for immediate profits led them away from agricultural diversity. If Middle St. John's planters maintained agricultural diversification through the Napoleonic Wars, as Chaplin argued in *An Anxious Pursuit*, the steady decline in long-staple cotton prices after 1819 motivated many to focus on agricultural specialization. Their financial and cultural welfare depended on producing higher yields of the long-fibered cash crop. "As long as cotton could be produced at a profit, however modest that profit might be," according to Smith, "the advantages of growing cotton seemed to be greater than those for any other crop." In 1846, the twenty-seven Biggin Basin plantations grew an average of eighty-two pounds per acre of cotton, with Wantoot growing up to 160 pounds per acre. The output averaged from 136 pounds per acre on the sea islands up to 325 pounds per acre in Florida. Yet, planters' decisions to focus solely on a single cash crop made them more susceptible to declines in price and quality.[59]

Santee long-staple planters began seeing their crop drop in quality by 1832. Henry William Ravenel observed in 1842 that, "the Santee cotton has lost that distinctive character it formerly possessed." Before this degeneration, "the average price of Santee cotton has generally been about 3 times that of Uplands [cotton]." Yet by the time of Ravenel's report, the Santee strain had fallen under this mark.[60] With declining cotton yields occurring in Middle St. John's, the increasing rate of decrement intensified the anxiety in planters.

[58] Porcher and Fick, *Sea Island Cotton*, 311; T.W. Peyre Journal, SCHS; "Account of an Agricultural Excursion," 78; Smith, *Economic Readjustment*, 59.

[59] Chaplin, *Anxious Pursuit*, 330, 345; Rosengarten, *Tombee*, 57–59; Smith, *Economic Readjustment*, 59; Peyre Journal; Gray, *Agriculture in Southern United States*, vol. 2, 737.

[60] "Agricultural Memoir," 131.

Out of desperation, some wealthy planters began to turn to alternative agricultural practices by the 1840s. Henry William Ravenel and his father, Henry, were the most vocal of those regional planters who believed in agricultural reform. Henry William Ravenel came from a family of planter-scientists. He was a distant cousin to Dr. Edmund Ravenel, a pioneering American conchologist. Edmund received a medical degree from the University of Pennsylvania in 1819 and was professor of chemistry and pharmacy at the Medical College of South Carolina. Resigning from his position in 1835, Edmund purchased the Grove Plantation on the Cooper River and used the property as a natural laboratory for scientific improvement. One of his interests was studying the geological formations of the region, uncovering the fossilized shells and marl along the Cooper River watershed. Edmund directed this study toward the early experimentation of fertilization of agricultural lands. A contemporary of Louis Agassiz, John Bachman, and John James Audubon, Edmund guided Sir Charles Lyell, the noted British geologist, through the Lowcountry in 1842 to assist in collecting geological specimens. Ravenel introduced Lyell to limestone outcroppings near the Santee River, which the geologist labeled "santee white limestone." Later the same year, Ravenel and Edmund Ruffin examined limestone deposits as a source for the state geological survey. The limestone deposits would become a focus of agricultural improvers in using marls and calcareous sediments as fertilizer.[61]

Dr. Henry Ravenel, Henry William's father, also pursued the planter-scientist vocation. Henry graduated from the New York College of Physicians and Surgeons and practiced medicine in Charleston. Henry's drive for scientific knowledge overlapped with his agricultural practices, as he started experimenting with soil renourishment at Pooshee Plantation. Inheriting the property from his father in 1827, Henry began using a system of husbandry to restore the land. Using manure acquired from horse stables as well as cattle, sheep, and hog pens, Henry incorporated the extract with leaves, cotton stalks, and grasses to form "compost." His efforts received "considerable and deserved success" from John Legaré in the *Southern Agriculturalist* and produced an average of 150 pounds of

[61] Lester D. Stephens, *Science, Race, and Religion in the American South: John Bachman and the Charleston Circle of Naturalists, 1815–1895* (Chapel Hill: University of North Carolina Press, 2000), 60–77; Albert E. Sanders and William D. Anderson, Jr., *Natural History Investigations in South Carolina: From Colonial Times to the Present* (Columbia: University of South Carolina Press, 1999), 59–61; Scarborough, "Science on the Plantation," 88–89.

cotton per acre by 1840. Despite Ravenel's attempts to restore the soil, his practices fell short of convertible husbandry. Henry refused to rotate fields, believing that the rewards were not worth the effort or the labor of his slaves to alternate subsistence crops with cotton.[62] Ravenel represented one of many "improving planters," according to environmental historian Steven Stoll, who encountered the "tangled negotiation between a desire for soil restoration through fodder crops and manure on one hand and the labor and land economics of planting on the other."[63]

Henry William Ravenel observed the scientific and agricultural practices set in motion by his father. Henry William's scientific upbringing contributed to him becoming one of the South's leading botanists of the Civil War era. He was born on his father's Pooshee Plantation in 1814. At eighteen years old, Henry William graduated from South Carolina College (now the University of South Carolina), where he studied chemistry in the hope of following his father's footsteps as a physician. His father discouraged the profession, according to Henry William, "as too laborious and liable to exposure for what he thought my weak constitution." He heeded his father's advice and took to planting, receiving Northampton Plantation from his father in 1835. By the time Henry William moved to the 800-acre property, Northampton had a poor reputation. Beginning with William Moultrie's failure to grow black-seed cotton, and continuing with a rapid succession of owners, local residents blamed poor soil and unskillful management for Northampton's failure. At age twenty-five, Henry William optimistically saw this plantation as an opportunity to integrate his scientific interests with plantation agriculture. Within seven years, Henry William had transformed the nutrient-depleted fields into a working model for husbandry. He expanded his father's fertilization practice by adding "plaster of Paris" (gypsum) to the compost mixture.[64] He followed Liebig's hypothesis that gypsum possessed one of several "essential nutrient elements" that plants needed to survive.

[62] Elise Pinckney, ed., "From the Collections: Dr. Henry Ravenel of Pooshee and Pineville," *South Carolina Historical Magazine* 81 (April 1980): 169–182; [John D. Legaré], "Account of the Management of Pushee, the Residence of Dr. Henry Ravenel; by the Editor," *Southern Agriculturalist and Register of Rural Affairs* 4 (July 1831): 360.

[63] Stoll, *Larding the Lean Earth*, 121.

[64] Quote: Arney R. Childs, ed., *The Private Journal of Henry William Ravenel, 1859–1887* (Columbia: University of South Carolina Press, 1947), 289–290; Haygood, *Henry William Ravenel*, 17, 20–21, 23–24; Sanders and Anderson, *Natural History Investigations in South Carolina*, 59; Richard D. Porcher and Douglas A. Rayner, *A Guide to the Wildflowers of South Carolina* (Columbia: University of South Carolina Press), 30–31; David Taylor, *South Carolina Naturalists: An Anthology, 1700–1860* (Columbia:

Instead of applying organic manures on fields, this new generation of scientific agriculturalists believed the solution was in medicating the land, or as Stoll notes "the inorganic theory like a doctor prescribing vitamins for nutrition instead of food."[65]

Ravenel's scientific experimentation with manures was a local solution to the statewide problem of mono-crop agriculture. Recognizing the relationship between soil depletion and cultural upheaval, cotton planters became concerned about the possibility of losing both economic and political power. As planters increased cotton production in response to falling prices, their slaves cleared more land and devoted more labor to cultivating the crop. As historian Theodore Rosengarten stated, "strained to produce the most year after year, the land would produce progressively less."[66] Lowcountry planters rightfully correlated emigration from South Carolina with less expensive and more fertile lands in the Deep South, as South Carolina lost approximately half of its white population to emigration after 1800. Planters could purchase nutrient-rich land more affordably than investing more labor to renourish declining cotton fields. Profit drove this movement, as planters simply did not see the economic advantage of investing labor, time, and resources into artificially sustainable land that they had previously decimated from overuse.[67]

South Carolinians saw a need for retaining potential planters due to the simple fact that many descendants of the agricultural elite were leaving the state in large numbers. Those planters who elected to stay felt fear and uncertainty from a demographic drain of an aspiring population and their enslaved laborers. State agricultural reformers also associated poor agricultural lands with their peers' growing anxiety of slavery and the loss of congressional representation in Washington, D.C. In the wake of the Nullification Crisis and political division over slavery, agricultural reformers saw new planting practices as one way of combating dwindling political power.[68]

University of South Carolina Press, 1998), 157–158; "Account of an Agricultural Excursion," 78; "Agricultural Memoir," 131.

65 Stoll, *Larding the Lean Earth*, 152–153.

66 Rosengarten, *Tombee*, 59.

67 James David Miller, *South by Southwest: Planter Emigration and Identity in the Slave South* (Charlottesville: University of Virginia Press, 2002), 19, 39–59.

68 For the connection between agricultural reform and nullification, see: Stoll, *Larding the Lean Earth*; Mathew, *Edmund Ruffin*; Miller, *South By Southwest*; Drew Gilpin Faust, *James Henry Hammond and the Old South: A Design for Mastery* (Baton Rouge: Louisiana State University Press, 1982); John Barnwell, *Love of Order: South Carolina's First Secession Crisis* (Chapel Hill: University of North Carolina Press, 1982).

One avenue to promote this knowledge with contemporaries was in regional and state agricultural societies. Through "addresses" or keynote speeches, South Carolinian planters revealed an array of prospects to stabilize fluctuating crop prices and promote new modes of agricultural land use. James Henry Hammond first presented the prospect of using inland rice for agricultural diversity in 1841. His argument was that South Carolinian planters must complement cotton production with an assortment of other crops to vary their income. Hammond noted how South Carolina grew three-quarters of the nation's rice crop in 1840, yet only a small percentage of the state's land was devoted to rice cultivation. Noting that rice could be grown up to two hundred miles inland and that there were "vast inland swamps well suited for it," Hammond encouraged midland and upcountry planters to consider the untapped wetlands suitable for rice cultivation. According to Hammond, inland rice cultivation would add to planters' wealth, relieve upcountry residents' dependence on Lowcountry rice, and improve the health of South Carolina by draining sickly swamps.[69]

Drew Gilpin Faust placed this speech into context by explaining that it was inspired by the 1839 decline in cotton prices, which Hammond assessed as a 33 percent decline in value.[70] Hammond saw agricultural diversity as a "rallying cry" for South Carolinians to combat emigration, plus provide economic and political stability. He believed changes in farming practices would generate more economic security for planters, which – in theory – would generate more national political power.[71] Rice was just one of several crops that Hammond believed farmers could grow to achieve agricultural and, thus, economic diversification. Tobacco and indigo, flax and hemp, Bene oil, viniculture, and silk were all products that planters could produce to complement their cotton crop. Ravenel recognized rice as an obvious crop for Biggin Basin planters to reintroduce in their former fields. Despite Ravenel's optimism toward growing rice, according to Rosengarten, he overlooked the "practical difficulties of supplanting the old staple such as building a network of brokers, buyers, and processors for cotton's replacement."[72]

69 James H. Hammond, 25 November 1841, in State Agricultural Society of South Carolina, *Proceedings of the Agricultural Convention and of the State Agricultural Society of South Carolina, from 1839 to 1845* (Columbia: Summer & Carroll, 1846), 184; J.R. Poinsett, "On Irrigation," *Southern Agriculturalist* 7 (November 1841): 614.

70 Faust, *James Henry Hammond*, 119.

71 Haygood, *Ravenel*, 30.

72 Rosengarten, "Southern Agriculturalist," 284.

Hammond and Ravenel's call for inland rice as a path to crop diversity rippled through the state and the Biggin Basin plantation community. As one reviewer of Hammond's speech stated in 1845, "we agree, therefore, with Gen. Hammond, in regarding rice as an important staple of which no competition is likely ever to deprive our State, and the sooner we enter into its more extensive cultivation, the better." Planters responded to this advice, as the state's upper districts increased rice production from less than 6,000 pounds in 1839 to more than 1 million pounds in 1849. In the middle districts, the crop expanded from slightly less than 1 million pounds to more than 3 million pounds in the ten-year period.[73]

At the time that Henry William Ravenel made his 1842 address, few Biggin Basin planters grew rice for profit. During an 1840 agricultural survey of St. John's, the editor of the *Southern Cabinet* noted that planters only grew "rice in sufficient quantities to supply the demand for family use." Notable cotton planter Thomas Walter Peyre's slaves annually planted an average one acre of rice for plantation consumption between 1835 and 1850. Henry Ravenel's half-brother, Thomas Porcher Ravenel, planted just two and a half acres of subsistence rice annually at Indianfield from 1846 to 1858, a plantation on which William Moultrie had devoted 126 acres to rice cultivation before the Revolution.[74] In his 1843 tour of the state, Edmund Ruffin noted that in the Biggin Swamp, "all rice culture ... has been abandoned, except for home consumption."[75] To Ruffin, inland rice farming was sickly in practice and he thought that draining inland swamps would improve the economic and physical health of the region. He believed that this fertile land, that once produced "the superior quality of the rice of the inland swamps," would produce large yields of cotton.[76]

Henry William Ravenel agreed with Ruffin's points on improving the inland rice environment, but he believed that rice planting best served these low-lying areas. Ravenel began the renewed interest of regional rice cultivation by planting on Northampton in 1839. By 1850, Ravenel produced a modest 3,000 pounds of clean rice on the property. By 1847, Peyre noted that thirteen out of thirty-one plantations in the area

[73] "The Agricultural Prospects of South Carolina: Her Resources and Her True Policy," *Southern Quarterly Review* 8 (July 1845): 118; Smith, *Economic Readjustment*, 67.

[74] "Account of an Agricultural Excursion," 13; T.W. Peyre Plantation Journal, SCHS, 18, 42, 67, 185, 193, 314; Thomas Porcher Ravenel Plantation Book, 1845–1854, Thomas P. Ravenel Papers, SCHS; "Plat of Indianfield Plantation," no. 4814, Plat Collection of John McCrady, CCRMC.

[75] Mathew, *Private Diary*, 72, 169.

[76] Ibid., 163; Mathew, "Agricultural Survey," 69–71.

were growing inland rice in "good" and "very good" quantities.[77] The planters of these thirteen plantations had a vested interest in rice culture, as they came from an earlier generation of planters. By 1850, Henry William Ravenel's father, Dr. Henry Ravenel, was growing rice on his properties. Although Henry William sold Northampton and moved to Aiken, South Carolina for health reasons in 1853, his father continued to cultivate rice, actually increasing his output from 3,500 pounds in 1850 to 13,000 pounds a decade later.[78]

Despite Middle St. John's planters' efforts to rejuvenate rice production in the 1840s, interest in the plant started to wane by 1850. Rising cotton prices motivated Middle St. John's planters to focus again on agricultural specialization. Of the thirteen plantations growing rice crops in Pyre's 1847 journal, only seven continued cultivation by 1850. Francis A. Porcher, at Somerton Plantation, grew 4,500 pounds of rice in 1850. Thomas F. Porcher grew just 1,200 pounds at Whitehall Plantation. The estate of Isaac Porcher grew 4,000 pounds at Chapel Hill and neighboring Moorfield Plantations. The estate of John P. Porcher, Jr. actively grew rice at Cedar Spring Plantation, with an impressive 45,000 pounds. Neighboring planter, Dr. Morton Waring, grew 7,500 pounds of rice at Chelsea Plantation. Finally, William James Dennis, former overseer of Edmund Ravenel, purchased Hog Swamp Plantation from his employer and grew 18,000 pounds of rice in 1850.[79]

While a core group of planters continued to cultivate the wetland crop into the 1850s, natural disasters took their toll on Middle St. John's inland rice cultivation. Freshets and droughts were prevalent in the region during that decade. Three hurricanes swept the South Carolina coastline between 1853 and 1857, with a major hurricane striking Charleston in 1854. The storm moved up the coast, sending destructive tidal surges and causing freshets that "broke through one or more rice dams on nearly every plantation" along the Cooper River, according to historian Walter

[77] Haygood, *Ravenel*, 26, 69; "1847 Middle St. John Plantation Inventory," T.W. Peyre Plantation Journal, SCHS, 300.

[78] Haygood, *Ravenel*, 68–70; 1850 South Carolina Agricultural Census, Charleston District – St. John's Parish, 367–368; 1860 South Carolina Agricultural Census, Charleston District – St. John's Parish, 345–346.

[79] "1847 Middle St. John Plantation inventory," T.W. Peyre Plantation Journal, SCHS, 300; 1850 South Carolina Agricultural Census, Charleston District – St. John's Parish, 367–368, 1860 South Carolina Agricultural Census, Charleston District – St. John's Parish, 345–346.

J. Fraser, Jr.[80] Even heavy rainstorms had their destructive tendencies in Biggin Basin. Thomas Porcher Ravenel described how "the whole country [was] under water" after thirteen inches fell between September 3 and 12. Ravenel painted a dim picture of drainage conditions, explaining in his diary that "places [were] entirely drowned." He observed that the "ditches remain[ed] full and streams continue[d] high from the saturation and sobbing of the earth."[81] While freshets washed Middle St. John's planters out, droughts in 1853 and 1855 dried up impounded reservoirs. Ravenel noted how a "bad drought" in 1853 evaporated all the reservoirs in the area "except for the largest springs." Two years later, a similar drought affected the basin, once again drying up all the "ponds."[82] To planters attempting to break from mono-crop practices, the natural disaster proved too much for their economic welfare. Managing to control freshets or siphon water during droughts proved an afterthought to the Biggin Basin planters whose primary concern was the success of their cotton crop.

Despite planters' interest in agricultural diversity, Middle St. John's inland rice collapsed by the eve of the Civil War. Cotton prices rose during the 1850's, tempting planters to focus all of their efforts once again on the cash crop. As Alfred Glaze Smith, Jr., observed, "if the chances of diversification of crop production had been limited previously they were even more so after 1850." Smith believed that the stimulus of increasing cotton prices and improvement in production conditions in the 1850s gave planters "an optimism which might be termed refreshing after the rather unrelieved pessimism which had pervaded the State during the 1840s." Middle St. John's planters adopted that optimism. After Dr. Henry Ravenel purchased Woodboo from the Mazyck family in 1859, his son Thomas Porcher Ravenel commented on clearing "the old canal from the Santee Canal to the spring with five men." He noted that the Woodboo canal had not been cleared out in "perhaps forty to fifty years" and it took these five people sixteen days to complete the task.[83]

[80] Walter J. Fraser, Jr., *Lowcountry Hurricanes: Three Centuries of Storms at Sea and Ashore* (Athens: University of Georgia Press, 2006), 119–134, quote: 129.

[81] September 1852, Thomas Porcher Ravenel Plantation Book, 1845–1854, Ravenel Family Papers, SCHS.

[82] Thomas Porcher Ravenel Plantation Book, 1845–1854 and 1855–1874, Ravenel Family Papers, SCHS, quotes: 1853 and 1855.

[83] Smith, *Economic Readjustment*, 76, 111; 12–17 September, 12–19 October, 22–26 November 1859, Thomas Porcher Ravenel Plantation Book, 1855–1874, Ravenel Family Papers, SCHS.

Even with the Mazyck family growing rice on the plantation, they failed to keep up the inland rice infrastructure.

Before agricultural reform passed through the Lowcountry in the 1840s, Middle St. John's planters viewed inland rice cultivation as a relic from the previous century. Yet with promotion by Henry William Ravenel, people reevaluated this older culture and incorporated the grain into their overall plantation economy. Inland rice culture was a symbol of success in the colonial era, while abandoned drains and embankments in antebellum Middle St. John's provided a constant reminder of past agricultural practices. Because of Biggin Basin's close association with inland rice planting, its mid-nineteenth century boosters saw it as an alternative crop to complement the unstable cotton market. Despite the limited efforts to revive rice cultivation in Middle St. John's, representing the limited effort of agricultural reform in general, planters abandoned this practice by the eve of the Civil War. The market promoted long-staple cotton once again. With the temptation of monetary gain, Middle St. John's planters shed any possibility of continuing their ancestral legacy of growing rice.

7

Epilogue

Forgotten Fields

The lasting results of the Civil War came from the emancipation of enslaved labor and the demise of the plantation system. The removal of slave labor led to a restructuring of economic and social patterns, with agricultural endeavors suffering a slow demise in the Lowcountry. African Americans could migrate freely and this was represented in the shifting labor populations in rural communities as many sought employment in urban centers or other rural communities. However, the two most successful cash crops in the antebellum Lowcountry – rice and cotton – did not disappear overnight. Instead, agricultural landscapes shifted over time as landowners failed to turn profits and compete with new technologies, and as they succumbed to natural disasters.[1]

Huger Creek planters struggled to maintain economic security from rice culture after the Civil War. Rice output on Limerick dropped from 558,830 pounds in 1860 to 2,000 pounds ten years later. William J. Ball did recover by bringing the output up to 24,000 pounds in 1880, but his economic success did not compare to before the war. The Gibbs family did not fare much better at Windsor. John C. Gibbs managed Windsor for his mother, but the family sold the property in tracts, which were purchased by land speculator Charles Greenland McCay in 1878 and Ada Guilds in 1886. The lack of labor, combined with increased labor costs, made large scale inland rice planting inefficient. With the collapse of the

[1] James H. Tuten, *Lowcountry Time and Tide: The Fall of the South Carolina Rice Kingdom* (Columbia: University of South Carolina Press, 2010).

inland rice culture, Huger Creek planters sought income from cattle ranching, cotton, corn, and peas.[2]

Emancipated slaves living on these properties originally made agreements with the Ball and Gibbs families to establish new working conditions after the war. Twenty-one people signed a contract with J.C. Gibbs to work on the Windsor fields and reside on the property. While working conditions did not vary much from antebellum task systems, freedmen did posses some power in bargaining for better wages. At Windsor and Limerick, field hands received wages in the form of one-third of the total crop, which was customary during the Reconstruction period. Labors worked the "two-day system," which ment two or three days spent in the rice fields, and then they devoted the rest of the week to subsistence farming at their homestead. William J. Ball expressed frustration about this system and the lack of output from his former slaves, stating that fifty-three people were completing the same amount of work previously accomplished by thirty-four people under the "old system."[3]

What emancipation revealed about inland rice cultivation was the close association that the regime had to enslaved labor. Unlike tidal rice agriculture, which continued albeit a slow demise until the early twentieth century, commercial inland rice cultivation all but ceased to exist after the Civil War. Although there are several reasons why commercial inland rice production ended in the Lowcountry, for the most part, after the Civil War, the driving reason was economics. Inland rice plantations were not as efficient as tidal plantations. Peter Coclanis estimates that tidal irrigation, with consistent access to water and more effective flooding to kill competing weeds, produced up to twice the amount of rice per field hand compared to inland cultivation. In the postbellum era, where planters were faced with dwindling profits from the shift to wage labor and increasing domestic competition from new producers in Louisiana,

[2] William Lees, *Limerick: Old and in the Way, Archeological Investigations at Limerick Plantation, Berkley County, South Carolina* (Columbia: Institute of Archaeology and Anthropology, 1980), 72–74; M.L. Walker, *Abstract of Title Covering Dorchester Land and Timber Co., Tract #2a, Berkeley County, South Carolina, Containing 40,960 Acres* (Charleston: United States Department of Agriculture, Forest Service, 1934), 459–469.

[3] March 17, 1866, "M.G. Gibbs with Freedmen & women," Records of the Field Offices for the State of South Carolina, Records of the Bureau of Refugees, Freedmen and Abandoned Lands (microfilm), National Archives, Washington, D.C; "Plantation Journal and Reminisces from May 1865 to [n.d.]," 24 March 1866, W.J. Ball Papers, South Carolina Historical Society (SCHS), Charleston, SC; Stephen Hahn, *Nation Under Our Feet: Black Political Struggles in the Rural South from Slavery to the Great Migration* (New York: Belknap Press, 2003), 347.

Arkansas, and Texas, inefficient inland rice production could not keep up with market forces.[4]

Contrary to white-owned commercial enterprises, African Americans did continue inland rice cultivation on small-scale and subsistence levels throughout the Lowcountry into the twentieth century. As rice historian James H. Tuten notes, "lowcountry freedmen went into the uncertain post-bellum with articulated goals."[5] A central goal for freedmen was to own land. For land redistribution, the South Carolina Land Commission was a radical avenue for former slaves and landless whites to obtain acreage after the Civil War. The Land Commission was set up by the Reconstruction Legislature to purchase land through a sinking fund from state taxes and then sell subdivided tracts at fair market prices. In the words of Union League president Francis L. Cardozo, the commission gave "the poor people the opportunity to become owners of the soil they cultivate."[6] Unique to South Carolina, the Land Commission purchased "worn out rice fields" and land "capable of cultivation" for freedmen to have the opportunity to participate in the market economy.[7]

Contrary to the well-intentioned beliefs of this organization, the land commission was troubled by corruption and fraud. Landowners bribed officials to purchase land at inflated prices and officials profited from land sales when making commissions from real estate agents. Land officials bought several former inland rice plantations throughout the Lowcountry. Beginning in 1870, land speculators purchased former plantations below market value and quickly turned around to sell them to the South Carolina Land Commission. For example, Dr. Lewis Schley served as a cover to protect the identity of his brother-in-law, William T. Wragg, the realtor who purchased Wythewood and Ararat Plantations for $10,000 and then sold the properties to the state for $35,980.[8]

Inland rice plantations became lucrative tracts for land speculators to sell to the state Land Commission because state officials thought they

[4] Peter A. Coclanis, *The Shadow of a Dream: Economic Life and Death in the South Carolina Low Country, 1670–1920* (New York: Oxford University Press, 1989), 96–97; Tuten, *Lowcountry Time and Tide*, 45.

[5] Tuten, *Lowcountry Time and Tide*, 30.

[6] Hahn, *Nation Under Our Feet*, 260.

[7] South Carolina General Assembly, *Report and Resolutions to the General Assembly of the State of South Carolina, at the Regular Session, 1872–1873* (Columbia: Republican Printing Co., 1873), 143–144.

[8] Carol K. Rothrock Bleser, *The Promised Land: The History of the South Carolina Land Commission, 1869–1890*, Tricentennial Studies 1 (Columbia: University of South Carolina Press, 1969), 62.

could quickly turn a profit on these tracts. These former inland rice plantations ranged between $1 and $5 per acre in Charleston County in 1883. In comparison, fine tidal rice fields fetched between $20 and $30 per acre in 1883.[9] Contrary to the perceived value of these plantations, historian Carol Rothrock Bleser notes the former inland rice fields were "worthless for providing small homesteads" because they required a large labor population to maintain the rice fields. Wythewood Plantation became an example of how the commission catered to the wealthy instead of to the poor, because the land transaction generated a quick profit for agents instead of providing fertile land for subsistence freedmen. Rothrock notes that, "heavy capital investment and a combination of proprietors would have needed to restore [Wythewood] to anything like its original value."[10] Despite challenges cultivating the inland rice fields, seventeen freedmen purchased tracts varying between 21 and 50 acres. Although the land was not conducive to small-scale farming because the land was hard to drain, former slaves of Wythewood continued the practice of cultivating the wetland areas. Two families, the Porchers and the Wrights, purchased tracts overlapping former rice field divisions to continue cultivating rice on the existing infrastructure until the turn of the century. In a report to the State Officers Board, the Land Commission representative stated optimistically that "a few persons have ventured to locate lots on the tract, and many promise to try their hands [in cultivating rice]."[11]

Despite the best intentions of the Land Commission to redistribute former plantations to freedmen, corruption and embezzlement brought down the Land Office. "The story ended as it had begun in 1860 with the restoration of the plantations," according to Bleser; "by 1890 much of the land commission holdings were concentrated in the hands of a few white families."[12] Former inland rice plantations reserved for African Americans began once again appearing under the ownership of a new generation of white landowners. By the turn of the century, the little rice cultivation occurring on these lands had disappeared. Southern land speculators and northern businessmen began buying large tracts of

[9] Bleser, *Promised Land*, 60; State Board of Agriculture of South Carolina, *South Carolina: Resources and Population, Institutions and Industries* (Charleston: Walker, Evans, and Cogswell, 1883), 66–67, 57.

[10] Bleser, *Promised Land*, 65.

[11] South Carolina General Assembly, Report to the State Officers Board, *Report and Resolutions*, 1872, 144.

[12] Bleser, *Promised Land*, 144.

"utterly worthless" land, in the words of land surveyor J.E. Green, and appropriating these tracts into timber plantations.[13]

Logging operations on the former inland rice plantations are an example of how people once again redefined value to the land. From the late nineteenth century to the turn of the twentieth, forest growth returned to abandoned rice and cotton fields. Abandoned inland rice fields returned to cypress and hardwood communities. During this period, and into the early twentieth century, individual property owners began selling their plantations to lumber companies. By 1906, E.P. Burton Lumber Company purchased Limerick, Windsor, Fishbrook, and surrounding plantations to amass 47,000 acres. A.C. Tuxbury Lumber Company bought Wythewood and nearby inland rice plantations Awendaw Barony, Cypress Pond, Irishtown, Mt. Pleasant, Dog Swamp, and Walnut Grove, among others, to consolidate over 42,000 acres. These timber companies changed the landscape once again. They extracted hardwoods from the former rice fields, while harvesting longleaf and loblolly pine in the upland areas. Instead of using enslaved labor to dig drainage canals and construct rice embankments, these timber companies built an infrastructure of railroads, causeways, and logging camps to extract highland timber quickly, leaving a clear-cut environment within a fifteen-year period.[14]

E.P. Burton cut in the Huger Creek watershed, while A.C. Tuxbury operated between the East Branch of the Cooper and the Wando rivers. Between 1899 and 1902, E.P. Burton operated a commissary, employee housing, and a blacksmith shop next to Huger Bridge. The village catered to the initial stages at Limerick and was connected to Burton's railroad by a spur branch. As E.P. Burton expanded into Windsor and Fishbrook, the company built a second village beginning in 1902. Located approximately five miles from the E.P. Burton dock at Silk Hope on the East Branch of the Cooper, Conifer originally consisted of employee housing and cook's quarters. In four years the village grew to a population of 500 people, supporting a blacksmith's shop, commissary, superintendent's office, doctor's office, and a company house to lodge the foresters. By 1906, E.P. Burton railroads crossed eleven miles of the former plantations. Under ideal conditions, employees could lay up to 150 yards of ties and rails a day. However, the varied terrain presented challenges to

[13] Ibid., 64.

[14] Al Hester, "Establishing the Francis Marion: National Forest History in South Carolina's Lowcountry, 1901–1936," *Forest History Today* 17 (Spring/Fall 2011): 56–58.

building log trestles and earthen causeways over the wetlands. Burton engineers either reinforced former rice embankments or constructed new earthworks to support the locomotives and cars encroaching into the forestlands. For example, they used a Windor rice field embankment as a logging causeway and the Fishbrook dam for their railroad causeway that connected Conifer with the Silk Hope dock.[15]

Logging activity within the Huger Creek watershed was a result of new technologies to extract timber, such as band saws and skidders, combined with the increased demand for wood products in Charleston and other Southern cities.[16] Still reeling from the economic collapse after the Civil War, Charleston and its port depended upon the growing logging industry taking place within the Huger Creek watershed. By 1913, three Charleston-area timber companies had a cumulative annual production of over 300 million board feet.[17] E.P. Burton deforested their Huger Creek tract by 1916, with both Dorchester Land and Timber Company and A.C. Tuxbury conducting a second cut by 1924. Timber companies cut over Limerick, Windsor, and Fishbrook Plantations to such an extent that land reformers viewed this spatial boundary as an ideal location to promote the idea that scientific forestry management could solve larger land problems occurring through the Southeastern Coastal Plain.[18]

As a result of timber companies rapidly depleting forests throughout the eastern United States, people began developing new interpretations of conservation of natural resources. Gifford Pinchot, father of the conservation movement, saw humans as stewards of the land but also believed that nature is meaningful only when it serves multiple and practical human purposes. To Pinchot, forestry was both an art and a science. By addressing expanding forestry issues, he believed that industrial logging could safely continue with the expertise of scientifically trained professionals. Pinchot carried this ideology with him to the newly created United States Division of Forestry, where he became Chief Forester in 1898. Two years later he helped created the Yale Forestry School, the nation's premier forestry education and research institution. Pinchot's conservation success was reflected in the area of land managed by the

[15] C.S. Chapman, *A Working Plan for Forest Lands in Berkeley County, South Carolina*, Bulletin 56 (Washington, D.C.: US Department of Agriculture, Bureau of Forestry, 1905), 19–22; Thomas Fetters, *Logging Railroads of South Carolina* (Forest Park, IL: Heimburger House Publishing Co., 1990), 1416.

[16] Ted Steinberg, *Down to Earth: Nature's Role in American History* (New York: Oxford University Press, 2002), 62–67, 110–111.

[17] Hester, "Francis Marion," 56. [18] Ibid., 58–59.

Division of Forestry, which grew by approximately 150 million acres during Theodore Roosevelt's tenure as President, from 1901–1909.[19]

To Pinchot, sustained yield of timber was the first part of a dual concept system for managing forestlands. He followed a business-like philosophy where forests represented capital and the annual growth of forests represented the interest. By practicing sustained yields, timber companies would harvest the interest, or the annual growth, of the forest. The second concept stressed cooperation between the Division of Forestry and private landholders. Pinchot believed that the federal government should consult timber companies by emphasizing scientific forestry and sustained yield practices.[20]

As part of this cooperative effort, forestry assistant Charles S. Chapman and five assistants worked with E.P. Burton from December 1902 to March 1903 examining the forest's health within the Huger Creek watershed. Chapman's 1905 report provided a snapshot into forestry practices and landscape alteration at the time. Chapman stressed the importance of healthy growth and harvesting practices of the loblolly pine stands, representing 34 percent of forested E.P. Burton land. Loblolly became the dominant species on abandoned upland fields, replacing cotton, corn, and peas. Loblolly also became the dominant species after a clear-cut of longleaf stands. Recognizing the economic benefit of loblolly, "being a tree of very rapid growth and being well suited to the locality," Chapman advised E.P. Burton to remove the species' threat from fire, protect immature pines acting as seed trees, and cut trees with a diameter of fourteen inches or greater.[21] At the same time, Yale students affiliated with the Forest Service began working with E.P. Burton between 1902 and 1906 on the Limerick Plantation Tract. In 1928, A.B. Recknagel and the Cornell School of Forestry conducted research on Limerick, Windsor, and Fishbrook. This property became an early platform for the development of North American silviculture, a precursor to the mission of the Francis Marion National Forest.[22]

The Francis Marion National Forest originated from a series of initiatives addressing the complexity of forest land-use and management. By the second decade of the twentieth century, Progressive era thought

[19] Char Miller, *Gifford Pinchot and the Making of Modern Environmentalism* (Washington, D.C.: Island Press, 2001), 117–118, 141–144, 220–221, 156–157.

[20] Ibid.

[21] Chapman, *Working Plan for Forest Lands*, 19–21.

[22] Hester, "Francis Marion," 57–58; A.B. Recknagel, "Remeasuring the Hell Hole Plots in South Carolina," *Journal of Forestry* 26 (1928): 823–824.

worked its way into South Carolina through a compilation of federal and state conservation efforts. A gradual movement of Forest Service officials recognized that cooperation with individual landowners and timber companies could not solve universal deforestation and land mismanagement. Instead, the Forest Service took steps to solve the land problem through direct federal land acquisition and land management. The Weeks Act of 1911 and the Clarke-McNary Act of 1924 resulted from this philosophy, granting the federal government permission to acquire lands specifically for timber production plus providing forest protection funds to states that had established forestry departments.[23]

The Huger Creek watershed became a model for restoration in the Lowcountry by 1927. E.P. Burton deforested their Huger Creek tract by 1916, with Dorchester and Tuxbury conducting a second cut by 1924. Timber companies clearcut Limerick, Windsor, and Fishbrook Plantations to such an extent that the U.S. Forest Service saw this location as an ideal location to practice scientific forestry principles. In turn, silviculture deemed successful on these old inland rice plantations, according to Forest Service officials, could provide a model for reforesting other tracts through the Southeastern Coastal Plain. The newly created South Carolina State Forestry Commission served as a mediator between the Forest Service and the timber industry, while state and federal agencies began identifying purchase units in 1927. The National Forest Reservation Commission established the Wambaw purchase unit in February 1928, yet five years passed before New Deal stimulus enabled the Forest Service to purchase individual tracts. Between 1933 and 1935, the Forest Service secured 195,000 acres, approximately 80 percent of the land within the current Francis Marion boundary. On July 10, 1936, Franklin D. Roosevelt proclaimed the former Wambaw unit as the Francis Marion National Forest. Composed primarily of old inland rice plantations, the Francis Marion provided a new role of forestry and silviculture in the twentieth and twenty-first centuries.[24]

Today, people have a mixed understanding of inland rice cultivation. Often the cultivation practice is lumped together with the more visible tidal systems. Other times, the inland rice process is overlooked entirely, lost in the thick reforested environments that enslaved people worked so

[23] Miller, *Gifford Pinchot*, 156–157, 284–286; P.J. Paxton, *The National Forests and Purchase Units of Region Eight* (Atlanta: US Department of Agriculture, Forest Service, 1950), 47; Hester, "Francis Marion," 59.

[24] Paxton, *National Forests*, 47; Hester, "Francis Marion," 59–62.

hard to alter. Abandoned inland rice reservoirs, according to author Henry Ravenel Sass, represent the most "beautiful places in the Low-Country, it is hard to believe that there are more beautiful places anywhere on earth, than the cypress lagoons in some of our swamps." He notes that, "though they appear perfectly wild today, [the reservoirs] were made by the old rice planters in the course of the extensive engineering operations necessary for the production of the great crops of rice."[25] The second- and third-growth forests and vegetation reclaiming the human-altered wetlands created a misperception of a "pristine" wilderness. Sass reflects on this reclamation, stating "today [inland reservoirs] are among the loveliest and most interesting places in the Low-Country, for in most cases nature has hidden all obtrusive signs of man's handiwork, while herons, ibises, water turkeys and other wild creatures have taken the reserves for their own and established their populations cities there." The legacy of select inland reserves is that the impoundments have established themselves, as per Sass's description, as a magnet for wading birds and nature-lovers alike. Stanyarne Hall and Laurel Hill Plantations (now Caw Caw County Park) in the Rantowles Basin hosts 230 species of birds. Wythewood rice fields and canals now form the attraction for the I'on Swamp Trail in Francis Marion National Forest, and Myrant's Reserve at Fairlawn Plantation was the location where birders "rediscovered" the Bachman's Warbler, thought to be extinct.[26]

With development encroaching further from Charleston and other coastal hamlets into the countryside, new generations of residents must make decisions on what kind of value the land holds. The growing population in the twenty-first century Lowcountry has turned former inland rice plantations into contested lands. On one hand, these large

[25] Herbert Ravenel Sass, "The Low-Country Lagoons," *Charleston Museum Quarterly* 2 (First Quarter 1932): 20. Sources addressing the analysis of vegetation growth being younger than previously interpreted on abandoned inland rice fields and reservoirs is found in Brooke V. James and Benjamin J. Collins, *Secondary Succession Patterns of a Southern Bottomland Hardwood Forest on Former Agricultural Lands in and around the Santee Experimental Forest, South Carolina* (Cordesville, SC: Santee Experimental Forest, USDA Forest Research Station, December 2010), 1–32; and David Chamberlain, "Additional Notes on Bachman's Warbler," *The Chat* 67 (Winter 2003): 5–10.

[26] Janice Shumake, "Nature Blends with Past," *Post and Courier*, January 29, 1998; Editorial, "Fairlawn Is a Win-Win-Win Green Deal for County," [Charleston] *Post and Courier*, 29 July 2012: A12. Rev. John Bachman identified the Bachman's Warbler on Fairlawn Plantation in 1833; Arthur T. Wayne, "The Nest and Eggs of Bachman's Warbler, *Helminthophila bachmani* (Aud.), Taken near Charleston, South Carolina," *The Auk* 24 (January 1907): 43–48.

tracts provide opportunity for commercial and residential development. With some timber and paper companies altering their business models away from logging and paper manufacturing, residential development of former timberlands provides a viable strategy in creating profits from large amount of acreage. For example, MeadWestvaco has drafted plants to transform 78,600 acres of the Rantowles Basin into a mixed-use community. Like early colonists, modern developers have seen the possibilities of upland areas for houses and office parks, while leaving the low-lying wetlands to develop at a later time. On the other hand, former rice plantations have become symbols for modern conservation efforts. Beginning with the establishment of Francis Marion National Forest in 1936 and continuing through public and private efforts, these former plantations represent sacred grounds for wilderness and historical preservation.[27]

Modern development and preservation of these former inland rice plantations reflects how our society attempts to use the land in relation to larger political and economic factors. In a sense, that is the history of inland rice cultivation for over three hundred years and it is a central theme of this narrative. Beginning with the initial settlement of Carolina, European colonists interpreted the diverse landscape in a variety of ways. Pinelands resembled accessible park-like environments that appealed to European's vision of a new Acadia. In contrast, low-lying hardwood bottomlands discouraged colonial development based on colonists' inability to access the landscape. Shifts in these landscapes took place with the emergence of the Carolina rice economy and colonists' understanding of how to cultivate the grain in wetlands. With the implementation of reservoir impounded water and irrigation devices, to move water efficiently on and off the fields, inland swamps became desirable tracts by the mid-eighteenth century. The advancement of tidal technology by the last quarter of the eighteenth century and into the nineteenth century changed people's perceptions of the inland landscape once again. The

[27] Schuyler Kroff and Bo Peterson, "North Charleston Wins Watson Hill Battle," *Post and Courier*, May 17, 2011; John P. McDermott, "The Whole Package: MeadWestvaco's 3-Year-Old Real Estate Unit Looks to Cover All Bases," *Post and Courier*, January 24, 2011; Lynne Langley, "Easement Protects Ravenwood," *Post and Courier*, February 10, 2002; Lynne Langley, "Conservation Easement to Safeguard Bluff Plantation," *Post and Courier*, October 23, 2001; Robert Behre, "Colleton Plantation on National Register," *Post and Courier*, May 19, 1997. For the intersection between former rice plantations and conservation, see: Virginia Christian Beach, *Rice and Ducks: The Surprising Convergence that Saves the Carolina Lowcountry* (Charleston: Evening Post Books, 2014).

plantations took on a secondary status, seen in legal interpretations and declining real estate values. By the antebellum era, with declining real estate values for inland rice plantations, these tracts were used as a more affordable way for merchants and professionals to enter into the elite planter circles.

Connected to these changing perceptions of landscape was inland rice planters' struggle with the natural world and how that struggle effected their agricultural decisions. As planters and slaves altered the low-lying environments to increase their agricultural output, they changed the form of these wetlands. Damming streams, constructing embankments, and digging canals disrupted the natural hydrology of inland floodplains. As a consequence, freshets intensified as seen in examples in Nicholson Creek and the Biggin Basin. On the other hand, planters' reliance on water intensified during droughts. With people moving onto the Carolina frontier during the eighteenth century, they brought new diseases, most notably malaria, and inadvertently created new environments for the pathogen to spread. Colonists' construction of reservoirs and rice field divisions created ideal conditions for anopheles mosquitoes to multiply. Finally, the growth of this mono-crop system contributed to declining soil fertility and pests, reflected in diminishing yields of inland rice by the eve of the Revolutionary War. These unintentional and unforeseen consequences played out in changing perceptions of inland rice production, seen in declining real estate values and abandonment of inland rice cultivation by the beginning of the antebellum era.

Intertwined in the story of inland rice, and the story of plantation agriculture in general, is the dependence on enslaved labor to carry out the planters' desires. Enslaved Africans and African Americans contributed to the development of inland rice plantations through physical labor and intellectual instruction. Slaves cleared an unimaginable amount of vegetation, moved a tremendous amount of earth, and constructed miles of embankments and ditches to establish the inland rice field system. At the same time, slaves contributed to a growing understanding of the subtleties of managing water through these watersheds and carrying out agricultural schedules to work within the natural boundaries of growing rice. Inland rice planters' dependence on enslaved labor became apparent after emancipation. While the tidal rice system continued in the South Carolina economy until the early twentieth century, inland rice cultivation all but collapsed after the Civil War. Only through small-scale farming by African Americans did inland rice cultivation continue in the Lowcountry until the turn of the twentieth century.

People's ability to cultivate inland rice in a variety of landscapes reflects the diversity of this cultivation practice. For over 300 years, people have irrigated rice fields from reservoirs and upland streams, and in a variety of environments. During the height of inland rice culture in the latter half of the colonial period, inland rice plantations existed along small-stream flood plains, in hardwood bottomlands, and next to brackish tidal river floodplains. This cultivation practice contrasted with the tidal irrigation system, which was limited to a very specific corridor along river floodplains close enough to the ocean to receive tidal waters, yet not too close to have brackish water kill the crop. For the inland system, the draw of fresh water from high land to low ground was the essential requirement to harness this resource into impounded rice fields.

What planters (and later scholars) initially perceived as a simple system for growing rice, inland rice cultivation actually became an intricate use of topography and agriculture. When dealing with limited water, people did not have the luxury of fully exploiting the natural resource. Planters and slaves had to creatively devise methods of using water to produce high yields of the cash crop. With this notion, inland rice development was not static. Plantation systems constantly evolved as planters accessed new landscapes, exploited available labor, and gained new technological understanding. As discussed, large-scale inland plantations developed simultaneously with tidal plantations in the colonial era and came to take on a similar aesthetic. Despite planters' investment of capital and labor into this infrastructure, the constant struggle of impounding water differentiated this society from their tidal counterparts. Perhaps this story of inland rice is a lesson in movement: the movement of free and enslaved people to new environments, the movement of water across manipulated landscapes, and the movement of knowledge to adapt to constantly changing natural and human forces.

Appendix

TABLE 1 *South Carolina inland rice output: yields per acre*

Year	Output/acre
1690	3 bushels
1712	20–40 bushels
1740	20 bushels
1791	33–40 bushels
1869	30–50 bushels

Sources: John Stewart, "Letters from John Stewart to William Dunlop," *South Carolina Historical and Genealogical Magazine* 32 (January 1931), 16; John Norris, *Profitable Advice for Rich and Poor, in a Dialogue, or Discourse between James Freeman, a Carolina Planter, and Simon Question, a West-Country Farmer, Containing a Description, or True Relation of South Carolina* (London: F. How, 1712), 40; John Oldmixon, *The British Empire in America: Containing the History of Discovery, Settlement, Progress and State of the British Colonies on the Continent and the Islands of Americas*, vol. I, 2nd ed. (London: J. Brotherton et al., 1741), 520; Lewis Gray, *History of Agriculture in the Southern United States to 1860* (Washington, D.C.: Carnegie Institution of Washington, 1933), vol. 2, 730; "A.W. Burnet to George K. Vernon, 22 February 1869," James Butler Campbell Papers, South Carolina Historical Society.

Bibliography

Primary Sources

Manuscripts

Berkeley County Register of Mesne Conveyance, Monks Corner, SC
Deed Books
British Manuscript Collections
Public Record Office: Colonial Office, America & West Indies (microfilm)
The Society for the Propagation of the Gospel in Foreign Parts (microfilm)
Charleston County Register Mesne Conveyance, Charleston, SC
Deed Books
John McCrady Plat Collection, 1680–1929
J.P. Gaillard Plat Collection
Plat Books, A–C
Charleston Library Society, Charleston, SC
John B. Irving, "Record of Windsor and Kensington Plantations, 1840–1888"
College of Charleston, Special Collections, Addlestone Library, Charleston, SC
John Cord Papers
Charles Cotesworth Pinckney Plantation Diary
Drayton Papers
Plowden Weston Papers
Wigfall Family Papers
Duke University, Manuscript Department, William R. Perkins Library, Durham, NC
John Archdale Papers
John Ball Papers
Keating Simons Ball Papers
Francis Corbin Papers
Samuel Cotes Letters

Eliza Lucas Pinckney Papers
Augustan Taveau Papers
Library of Congress, Washington, DC
Geography and Map Division. www.loc.gov/rr/geogmap/
Middleton Place, SC
Middleton Family Papers
National Archives, Washington, DC
Records of the Bureau of Refugees, Freedmen and Abandoned Lands (microfilm)
South Carolina Department of Archives and History, Columbia, SC
Census Records, 1790–1930
Colonial Land Grants, 1688–1799
Colonial Plat Books, 1731–1775
Commissioners of Fortified Estates, 1782–1785
Committee Reports, 1782–1866
Conveyance Books, 1719–1775
Individual Tax Returns for 1824
Journal of the Commons House of Assembly of South Carolina, 1695–1775
Judgment Rolls, Court of Common Pleas, 1703–1839
Miscellaneous Inventories and Wills, Charleston County, 1687–1785
Petitions to the General Assembly, 1782–1866
Records in the British Public Record Office Relating to South Carolina, 1663–1782
Records of the Court of Chancery of South Carolina, 1671–1779
Resolutions of the General Assembly, 1782–1865
South Carolina Estate Inventories
South Carolina Memorials, 1731–1776
State Plat Books, 1784–1860
Warrants for Lands in South Carolina, 1672–1711
Wills of Charleston County, 1671–1868
South Carolina Department of Natural Resources, Geographic Information System (GIS) Division.
South Carolina Historical Society, Charleston, SC
Baker Family Papers
Ball Family Papers
W.J. Ball Papers
Sandford William Barker Papers
Black Oak Agricultural Society, Minutes
Bluff Plantation Book, 1759–1774
William Cain Papers
James Butler Campbell Papers
Cheves-Middleton Papers
Isaac M. Dwight Family Papers
Gibbes Family History and Genealogy Research Files
Mathurin Guerin Gibbs Plantation Register

Heyward Family Papers
Huger Family Papers
Dr. John Beaufain Irving Family Papers
Francis Kinloch Papers
Peter Manigault Papers
Maps and Muniments Series
Maps and Plats File
Henry A. Middleton Papers
Thomas Middleton Papers
Thomas Milliken Papers
T.W. Peyre Plantation Journal
Pinckney Family Papers
Porcher Family Papers
Robert Pringle Letterbook
Ravenel Family Papers
Ravenel Land Papers
Thomas P. Ravenel Papers
Benjamin Huger Rutledge Papers
Weehaw Plantation Miscellaneous File

University of North Carolina, Southern Historical Collection, Chapel Hill, NC

John Ball and Keating Simons Ball Books, 1779–1911
William J. Ball Books
British Public Records, ca.1600–1782
William Butler, "Observations on Rice Culture"
John Berkeley Grimball Papers
Heyward-Ferguson Papers
Manigault Family Papers
Arthur Middleton Papers
Thomas Middleton Plantation Book
North Carolina Map Collection. www.lib.unc.edu/dc/ncmaps
Josiah Smith, Jr., Lettercopy Book
Francis Stewart Papers

University of South Carolina, South Caroliniana Library, Columbia, SC

Ball Family Papers
Ball and Gilchrist Family Papers
James Glenn Papers
Manigault Family Papers
Mazyck Family Papers
Mulberry Plantation Journals, 1853–1889
Phillip Porcher Papers
Henry William Ravenel Papers
Keating Simons Papers
Taylor Family Papers
Louis Thibou Papers
Thomas Family Papers

Newspapers and Journals

Charleston Courier (Charleston, SC)
Charleston Mercury (Charleston, SC)
City Gazette (Charleston, SC)
DeBow's Review (New Orleans, LA)
Gentleman's Magazine (London, England)
News and Courier (Charleston, SC)
Post and Courier (Charleston, SC)
Russell's Magazine (Charleston, SC)
South Carolina and American General Gazette (Charleston, SC)
South Carolina Gazette (Charleston, SC)
South Carolina Gazette and Country Journal (Charleston, SC)
South Carolina State Gazette (Charleston, SC)
Southern Agriculturalist (Charleston, SC)
Southern Agriculturalist and Register of Rural Affairs (Charleston, SC)
Southern Cabinet of Agriculture, Horticulture, Rural and Domestic Economy (Charleston, SC)
Southern Cultivator (Augusta, GA)
Southern Patriot (Charleston, SC)
Southern Quarterly Review (Columbia, SC)

United States Census

South Carolina, 1790, 1800, 1820, 1830, 1840, 1860, 1870, 1880
South Carolina Agriculture Census, 1850, 1860, 1870, 1880

South Carolina Slave Census, 1850, 1860 Books and Articles

Allston, Robert F.W. *Memoir of the Introduction and Planting of Rice in South Carolina*. Charleston: Miller & Brown, 1843.

Archdale, John. *A New Description of That Fertile and Pleasant Province of Carolina*. London, 1707; Charleston: A.E. Miller, 1822.

Barton, William. "Observations on the Probabilities of the Duration of Human Life, and the Progress of Population, in the United States of America: In a Letter from William Barton, Esq. to David Rittenhouse, L.L.D. President, A.P.S." *Transactions of the American Philosophical Society* 3 (1793): 25–62.

Boswell, George. *A Treatise on Watering Meadows*. London: Printed for the Author, and Sold by J. Almon, 1779.

Carolina, Described More Fully Then Heretofore. Dublin: 1684.

Chalmers, Lionel. *An Account of the Weather and Diseases of South-Carolina*. London: Printed for Edward and Charles Dilly, 1776.

Cheves, Langdon, ed. *The Shaftesbury Papers and Other Records Relating to Carolina*. Charleston: South Carolina Historical Society, 1897; reprint, Charleston: Tempus Publishing, Inc., 2000.

Childs, Arney R., ed. *The Private Journal of Henry William Ravenel, 1859–1887*. Columbia: University of South Carolina Press, 1947.

Rice Planter and Sportsman: The Recollections of J. Motte Alston, 1821–1909. Columbia: University of South Carolina Press, 1953.

Clifton, James M., ed. *Life and Labor on Argyle Island: Letters and Documents of a Savannah River Rice Plantation, 1833–1867*. Savannah, GA: The Beehive Press, 1978.

Cohen, Robert, and Myriam Yardeni, eds. "Un Suisse en Caroline du Sud à la fin du XVIIe siècle," trans. by Harriott Cheves Leland, *Bulletin de la Société de l'Histoire du Protestantisme Français* 134 (1988): 59–71.

Cooper, Thomas, and David J. McCord, eds. *The Statutes at Large of South Carolina*. 10 vols. Columbia: A.S. Johnston, 1836–1841.

Crouse, Maurice A., ed. "The Letterbook of Peter Manigault, 1763–1773." *South Carolina Magazine* 70 (April 1969): 79–96; (July 1969): 177–195.

DuBose, Samuel. *Reminisces of St. Stephen's Parish, Craven County and Notices of Her Old Homesteads*. Charleston: A.E. Miller, 1858.

Address Delivered at the 17th Anniversary of the Black Oak Agricultural Society, April 27, 1858. Charleston: A.E. Miller, 1858.

Easterby, J.H., ed. *The South Carolina Rice Plantation, as Revealed in the Papers of Robert F. W. Allston*. Chicago: University of Chicago Press, 1945.

Ford, Timothy. "Diary of Timothy Ford, 1785–1786. With Notes by Joseph W. Barnwell." *South Carolina Historical and Geological Magazine* 13 (July 1912): 132–147; (October 1912): 181–204.

Glen, James. *A Description of South Carolina: Containing Many Curious and Interesting Particulars Relating to the Civil, Natural and Commercial History of That Colony in Colonial South Carolina: Two Contemporary Descriptions*. Chapman J. Milling, ed. Columbia: University of South Carolina, 1951.

Hall, Cpt. Basil. *Travels in North America, in the Years 1827 and 1828*. 2 vols. Philadelphia: Carey, Lea, and Carey, 1829.

Harper, Francis, ed. *The Travels of William Bartram: Naturalist's Edition*. New Haven: Yale University Press, 1958; reprint, Athens: University of Georgia Press, 1998.

Hewatt, Alexander. *An Historical Account of the Rise and Progress of the Colonies of South Carolina and Georgia*. 2 vols. London: Alexander Donaldson, 1779.

Hollis, Margaret Belser, and Allen H. Stokes, ed. *Twilight on the South Carolina Rice Fields: Letters from the Heyward Family, 1862–1871*. Columbia: University of South Carolina Press, 2010.

Holms, F.S. "Notes on the Geology of Charleston, SC." *American Journal of Science and Arts* 7 (March 1849): 187–201.

Irving, John B. *Day on the Cooper River*. Charleston: A.E. Miller, 1842.

Laurens, Henry. "An Original Letter of Henry Laurens." *Historical Magazine* 10 (April 1866): 99.

Lockwood, Thomas P. *A Geography of South Carolina*. Charleston: J.S. Burges, 1832.

Mathew, William M., ed. *Agriculture, Geology, and Society in Antebellum South Carolina: The Private Diary of Edmund Ruffin, 1843*. Athens: University of Georgia Press, 1992.

Matthews, Maurice. "A Contemporary View of Carolina in 1680." *South Carolina Historical Magazine* 55 (July 1954): 153–159.

Merrens, H. Roy, ed. "A View of Coastal South Carolina in 1778: The Journal of Ebenezer Hazard." *South Carolina Historical Magazine* 73 (October 1972): 177–193.

The Colonial South Carolina Scene: Contemporary Views, 1697–1774. Tricentenial Edition 7. Columbia: University of South Carolina Press, 1977.

Milligen-Johnson, George. *A Short Description of the Province of South Carolina, with an Account of the Air, Weather, and Diseases, at Charles Town, Written in the Year 1763*. London: Printed for John Hilton, 1770.

Morse, Jedidiah. *The American Universal Geography*. Boston: Thomas & Andrews, 1802.

Myrick, Jordan. "On Rice." In *Letters and Extracts on Agriculture, Published by the Order of the Agricultural Society of South Carolina*. Charleston: A.E. Miller, 1824.

Norris, John. *Profitable Advice for Rich and Poor, in a Dialogue, or Discourse between James Freeman, a Carolina Planter, and Simon Question, a West-Country Farmer, Containing a Description, or True Relation of South Carolina*. London: F. How, 1712.

Oldmixon, John. *The British Empire in America: Containing the History of Discovery, Settlement, Progress and State of the British Colonies on the Continent and the Islands of Americas*. Vol. I, 2nd ed. London: J. Brotherton et al., 1741.

Olmsted, Fredric Law. *A Journey in the Seaboard Slave States, with Remarks on Their Economy*. New York: Dix and Edwards, 1856.

Pinckney, Charles Cotesworth. "Two Letters from Charles Cotesworth Pinckney to Ralph Izard." *South Carolina Historical and Genealogical Magazine* 21 (October 1940): 150–152.

Pinckney, Elise, ed. "From the Collections: Dr. Henry Ravenel of Pooshee and Pineville." *South Carolina Historical Magazine* 81 (April 1980): 169–182.

Poinsett, J.R. "On the Draining Improvement of Rice Fields." *Proceedings of the Winyaw and All Saints Agricultural Society, April 18, 1850*. Charleston: Burgess, James, and Paxton, 1850.

Porcher, Frederick A. "Upper Beat of South Carolina: A Memoir." *Transactions of the Huguenot Society of South Carolina* 13 (1906): 1–48.

Ravenel, Edmund. "The Limestone Springs of St. John's Berkeley." *Proceedings of the Elliott Society* (October 1860): 28–31.

Ravenel, Henry Edmund, ed. *Ravenel Records: A History and Genealogy of the Huguenot Family of Ravenel, of South Carolina*. Atlanta: Franklin Printing and Publishing, 1898.

Richardson, J.S.G. *Reports of Cases at Law: Argued and Determined in the Court of Appeals and Court of Errors of South Carolina*. 15 vols. Columbia: A.S. Johnson, 1845–1869.

Robbins, Walter L., ed. "John Tobler's Description of South Carolina (1753)." *South Carolina Historical Magazine* 71 (July 1970): 141–161.

Salley, Alexander S., Jr., ed. *Narratives of Early Carolina, 1650–1708*. New York: Charles Scribner's Sons, 1911.

Simons, Thomas Y., M.D. "Remarks on the Climate of the Lower Country of South Carolina." *American Journal of the Medical Sciences* 9 (November 1831): 256.

Smith, J.L. "Analysis of the Cotton Lands on the Head Waters of the Cooper River." In *Report on the Geology of South Carolina*, M. Toumey, xliii–xlvii. Columbia: A.S. Johnston, 1848.

State Agricultural Society of South Carolina. *Proceedings of the Agricultural Convention and of the State Agricultural Society of South Carolina, From 1839 to 1845*. Columbia: Summer & Carroll, 1846.

Stewart, John. "Letters from John Stewart to William Dunlop." *South Carolina Historical and Genealogical Magazine* 32 (January 1931): 1–33; (April 1931): 81–114; (July 1931): 170–174.

Stony, Samuel Gaillard, ed. "Memoirs of Frederick Augustus Porcher." *South Carolina Historical and Genealogical Magazine* 44–45 (April 1943): 65–80; (July 1943): 135–147; (October 1943): 212–219; (January 1944): 30–40; (April 1944): 80–98.

Strobhart, James A. *Reports of Cases Argued and Determined in the Court of Appeals and Court of Errors of South-Carolina, on Appeals from the Courts of Law*. 5 vols. Charleston: Walker & Burke, 1847–1851.

Sullivan, Buddy, ed. *Darien Journal of John Girardeau Legare, Ricegrower*. Athens: University of Georgia Press, 2010.

Tanner, Henry S. *A Brief Description of the Canals and Railroads of the United States*. Philadelphia: Published by the Author, 1834.

Webber, Mabel L., ed. "Col. Senf's Account of the Santee Canal." *South Carolina Historical and Genealogical Magazine* 28 (January 1927): 8–21; (April 1927): 112–131.

Government Reports

Abbot, Frederic V. *Preliminary Examination of Owendaw and Wando Rivers and Other Waters and Water Routes Connecting Bull's Bay and the Harbor of Charleston, South Carolina*. In U.S. Congress, House of Representatives, 51st Cong., 1st sess., Ex. Document 82, H. Doc. Washington, D.C.: GPO, 1890.

Agha, Andrew, Charles F. Philips, Jr., and Joshua Fletcher. *Inland Swamp Rice Context, c. 1690–1783*. National Register of Historic Places Nomination Form. National Parks Service, Department of the Interior, Washington, D.C., 2011. nationalregister.sc.gov/SurveyReports/HC08003.pdf

Anderson, David, and Patricia A. Logan. *Francis Marion National Forest Cultural Resources Overview*. Columbia: United States Department of Agriculture, Forest Service, 1981.

Aucott, Walter R. "*The Predevelopment Ground-Water Flow System and Hydraulic Characteristics of the Coastal Plain Aquifers of South Carolina*."

United States Geological Survey Water Investigations Report B6–4347. Washington D.C.: GPO, 1988.

Austin, Amory. *Rice: Its Cultivation, Production, and Distribution in the United States and Foreign Countries*. United States Department of Agriculture, Division of Statistics Report 6. Washington, D.C.: GPO, 1893.

Bailey, Ralph, Jr., Andrew Agha, and Charles F. Philips, Jr. *Cultural Resources Survey of Part One of the Country Road Environmental Analysis, Francis Marion National Forest, Berkeley and Charleston Counties, South Carolina*. Francis Marion and Sumter National Forests Cultural Resource Management Report 05-01. Columbia: United States Department of Agriculture, Forest Service, 2008.

Beauchamp, Keith H. "A History of Drainage and Drainage Methods." In *Farm Drainage in the United States: History, Status, and Prospects*. Miscellaneous Publication 1455. George A. Pavelis, ed., Washington, D.C.: United States Department of Agriculture, 1987.

Cable, John S. et al. *An Archeological Survey of 3,720 Acres in The Bethera Area, Wambaw and Witherbee Districts, Francis Marion National Forest*. Francis Marion National Forest Indefinite Services Survey Report 2. Columbia: United States Department of Agriculture, Forest Service, 1991.

Chapman, C.S. *A Working Plan for Forest Lands in Berkeley County, South Carolina*. Bulletin 56. Washington, D.C.: United States Department of Agriculture, Bureau of Forestry, 1905.

Cooke, C. Wythe. *Geology of the Coastal Plain of South Carolina*. Bulletin 867. Washington D.C.: United States Geological Survey, 1936.

Doar, W.R. III, and Ralph H. Willoughby. *Revision of the Pleistocene Dorchester and Summerville Scarps, the Inland Limits of the Penholoway Terrace, Central South Carolina*. Columbia: South Carolina Department of Natural Resources, Geological Survey, 2006.

Eidson, Jennie P., et al. *Development of a 10- and 12-Digit Hydrologic Unit Code Numbering System for South Carolina, 2005*. Washington D.C.: United States Department of Agriculture, 2005.

James, Brooke V., and Benjamin J. Collins. *Secondary Succession Patterns of a Southern Bottomland Hardwood Forest on Former Agricultural Lands in and around the Santee Experimental Forest, South Carolina*. Cordesville, SC: Santee Experimental Forest, United States Department of Agriculture, Forest Service Southern Research Station, 2010.

Knapp, S.A. "Rice Culture." *Farmers' Bulletin* 417. United States Department of Agriculture: Washington, 1910.

Latimer, W.J., ed. "Soil Survey of Berkeley County, South Carolina." In *Field Operations of the Bureau of Soils, 1916*. M. Whitney, ed., 483–520. Washington, D.C.: Bureau of Soils, 1921.

Lukesh, G. R. *Preliminary Examination of Waterway from Charleston, SC, to the North Santee River*. In U.S. Congress, House of Representatives, 66th Cong., 2nd sess., H. Doc. 524. Washington, D.C.: GPO, 1920.

McCartan, Lucy, E.M. Lemon, Jr., and R.E. Weems. *Geologic Map of the Area between Charleston and Orangeburg, South Carolina*, Map I-1472. Washington, D.C.: United States Geological Survey, 1984.

Meanley, Brooke. *Blackbirds and the Southern Rice Crop, Resource Publication 100*, Washington D.C.: United States Department of the Interior, Fish and Wildlife Service, 1971.

Paxton, P.J. *The National Forests and Purchase Units of Region Eight.* Atlanta: United States Department of Agriculture, Forest Service, 1950.

South Carolina General Assembly. *Report and Resolutions of the General Assembly of the State of South Carolina at the Regular Session, 1872–1873.* Columbia: Republican Printing Co., 1873.

State Board of Agriculture of South Carolina. *South Carolina: Resources and Population, Institutions and Industries.* Charleston: Walker, Evans, and Cogswell, 1883.

Taveau, Augustin L. "Rice Culture." In *Report to the Commissioner of Agriculture for the Year 1867.* 174–178. United States Department of Agriculture. Washington, D.C.: GPO, 1868.

United States Department of Agriculture. *Soil Survey of Charleston County, South Carolina.* Washington, D.C.: GPO, 1971.

Soil Survey of Beaufort and Jasper Counties. Washington, D.C.: GPO, 1980.

Soil Survey of Berkeley County, South Carolina. Washington, D.C.: GPO, 1980.

Soil Survey of Dorchester County, South Carolina. Washington, D.C.: United States Department of Agriculture, 1990.

United States Geological Survey. *The National Map.* Washington, D.C. http://nationalmap.gov/

Walker, M.L. *Abstract of Title Covering A.C. Lumber Company Tracts #1 and #1–II, Charleston County, South Carolina, Containing 18,530 Acres*, vol. II. Charleston: United States Department of Agriculture, Forest Service, 1934.

Abstract of Title Covering Dorchester Land and Timber Co., Tract #2a, Berkeley County, South Carolina, Containing 40,960 Acres. Charleston: United States Department of Agriculture, Forest Service, 1934.

Weems, Robert E., and Earl M. Lemon, Jr., *Geology of the Bethera, Cordesville, Huger, and Kittredge Quadrangles, Berkeley County, South Carolina.* Washington, D.C.: United States Geological Survey, 1989.

Wheaton, Thomas R., Mary Beth Reed, John S. Cable, Theresa Hamby, and Leslie Raymer. *Archaeological Site Testing of Willow Hall and Walnut Grove Plantations, Francis Marion National Forest.* Francis Marion and Sumter National Forests Cultural Resource Management Report 92-13. Stone Mountain, GA: New South Associates, 22 April 1992.

Williams, G. Ishmael, John S. Cable, Mary Beth Reed, Cindy Abrams, and Theresa M. Hamby. *An Archaeological Survey of 3,438 Acres in the Coastal Area, Wambaw & Witherbee Districts, Francis Marion National Forest.* Francis Marion and Sumter National Forests Cultural Resource Management Report 91-45. Columbia: United States Department of Agriculture, Forest Service, 1993.

Willoughby, Ralph H., and W.R. Doar, III. "Solution to the 'Two-Talbot' Problem of Maritime Pleistocene Terraces in South Carolina." Poster. Columbia: South Carolina Department of Natural Resources, Geological Survey, 2006.

Wood, Karen G. *Site Evaluation on Three Sites at Historic Clayfield Plantation, Wambaw Ranger District, Francis Marion National Forest, South Carolina.* Francis Marion and Sumter National Forests Cultural Resource Management Report 91-30. Columbia: United States Department of Agriculture, Forest Service, 1991.

Secondary Sources

Books and Articles

Alpern, Stanley B. "Did Enslaved Africans Spark South Carolina's Eighteenth Century Rice Boom?" In *African Ethnobotany in the Americas*. Robert Voeks and John Rashford, eds., 35–66. New York: Springer, 2013.

Armstrong, Catherine. *Landscape and Identity in North America's Southern Colonies from 1660 to 1745*. London: Ashgate, 2013.

Aucott, Walter R., and Gary K. Speiran. "Ground-Water Flow in the Coastal Plain Aquifers of South Carolina." *Ground Water* 23 (November/December 1985): 736–745.

Bailey, N. Louise, and Elizabeth Ivey Cooper, eds. *Biographical Directory of the South Carolina House of Representatives*. Volume III: *1775–1790*. Columbia: University of South Carolina Press, 1981.

Bailey, N. Louise, and Walter Edgar, eds. *Biographical Directory of the South Carolina House of Representatives*. Volume II: *The Commons House of Assembly, 1692–1775*. Columbia: University of South Carolina Press, 1977.

Bailey, N. Louise, Mary L. Morgan, and Carolyn R. Taylor. *Biographical Directory of the South Carolina Senate, 1776–198*. Volume II. Columbia: University of South Carolina Press, 1986.

Bailyn, Bernard. *The Peopling of British North America: An Introduction*. New York: Alfred A. Knopf, Inc., 1986.

Baldwin, Agnes Leland. *First Settlers of South Carolina, 1670–1700*. Easley, SC: Southern Historical Press, 1985.

Ball, Edward. *Slaves in the Family*. New York: Farrar Straus & Giroux, 1998.

Barnwell, John. *Love of Order: South Carolina's First Secession Crisis*. Chapel Hill: University of North Carolina Press, 1982.

Barry, John M. *Natural Vegetation of South Carolina*. Columbia: University of South Carolina Press, 1980.

Beach, Virginia Christian. *Rice and Ducks: The Surprising Convergence That Saves the Carolina Lowcountry*. Charleston: Evening Post Books, 2014.

Berlin, Ira. "Time, Space, and the Evolution of Afro-American Society on British Mainland North America." *American Historical Review* 85 (Feburary 1980): 44–78.

Billings, Warren M. "Law in the Colonial South." *Journal of Southern History* 73 (August 2007): 602–616.

Blassingame, John W. *The Slave Community: Plantation Life in the Antebellum South*. New York: Oxford University Press, 1972.

Bleser, Carol K. Rothrock. *The Promised Land: The History of the South Carolina Land Commission, 1869–1890*. Tricentennial Studies 1. Columbia: University of South Carolina Press, 1969.

Braudel, Fernand. *Mediterranean and the Mediterranean World in the Age of Philip II*. 2 vols. New York: Harper and Row, 1972.

Bray, Francesca. *The Rice Economies: Technology and Development in Asian Societies*. Oxford: Basil Blackwell, 1989; reprint, Berkeley: University of California Press, 1994.

Bray, Francesca, Peter Coclanis, and Edda L. Fields-Black, and Dagmar Schafer, eds. *Rice: A Global History*. New York: Cambridge University Press, 2015.

Brittain, John LaFayette. "Two Recently Discovered Letters of Charles Cotesworth Pinckney: Another Glimpse into the Mind of an Eighteenth Century Man of Affairs." *South Carolina Historical Magazine* 76 (January 1975): 12–20.

Bull, H.D. "Kinloch of South Carolina." *South Carolina and Genealogical Magazine* 46 (April 1945): 63–69.

Bull, Kinloch. "Barbadian Settlers in Early Carolina: Historical Notes." *South Carolina Historical Magazine* 96 (October 1995): 329–339.

Canady, Hoyt P. *Gentlemen of the Bar: Lawyers in Colonial South Carolina*. New York: Garland Publishers, 1987.

Carney, Judith. "From Hands to Tutors: African Expertise in the South Carolina Rice Economy." *Agricultural History* 67 (Summer 1993): 1–30.

"Landscapes of Technology Transfer: Rice Cultivation and African Continuities." *Technology and Culture* 37 (Spring 1996): 5–35.

"Rice Milling, Gender and Slave Labour in Colonial South Carolina." *Past and Present* 153 (November 1996): 108–134.

"The Role of African Rice and Slaves in the History of Rice Cultivation in the Americas." *Human Ecology: An Interdisciplinary Study* 26 (December 1998): 525–545.

Black Rice: The African Origins of Rice Cultivation in the Americas. Cambridge, MA: Harvard University Press, 2001.

Carney, Judith A., and Richard Nicholas Rosomoff. *In the Shadow of Slavery: Africa's Botanical Legacy in the Atlantic World*. Berkeley: University of California Press, 2009.

Carney, Judith, and Richard Porcher. "Geographies of the Past: Rice, Slaves, and Technological Transfer in South Carolina." *Southeastern Geographer* 33 (November 1993): 127–147.

Catterall, Douglas. "The Worlds of John Rose: A Northeastern Scot's Career in the British Atlantic World, c.1740–1800." In *A Global Clan: Scottish Migrant Networks and Identities since the Eighteenth Century*. Angela McCarthy, ed., 67–94. New York: Tauris Academic Studies, 2006.

Chamberlain, David. "Additional Notes on Bachman's Warbler," *The Chat* 67 (Winter 2003): 5–10.

Chaplin, Joyce E. "Tidal Rice Cultivation and the Problem of Slavery in South Carolina and Georgia, 1760–1815. *William and Mary Quarterly* 49 (January 1992): 29–61.

An Anxious Pursuit: Agricultural Innovation and Modernity in the Lower South, 1730–1815. Chapel Hill: University of North Carolina Press, 1993.

Childs, St. Julien Ravenel. *Malaria and the Colonization of the Lowcountry, 1526–1696*. Baltimore: Johns Hopkins University Press, 1940.

Clifton, James M. "The Rice Industry in Colonial South Carolina." *Agricultural History* 55 (July 1981) 266–283.

"The Rice Driver: His Role in Slave Management." *South Carolina Historical Magazine* 82 (October 1981): 331–353.

Clowse, Converse D. *Economic Beginnings in Colonial South Carolina 1670–1730*. Columbia: University of South Carolina Press, 1971.

Coclanis, Peter A. "Rice Prices in the 1720s and the Evolution of the South Carolina Economy." *Journal of Southern History* 48 (November 1982): 531–544.

The Shadow of a Dream: Economic Life and Death in the South Carolina Low Country, 1670–1920. New York: Oxford University Press, 1989.

"Thickening Description: William Washington's Queries on Rice." *Agricultural History* 64 (Summer 1990): 9–16.

"Distant Thunder: The Creation of a World Market in Rice and the Transformations It Wrought." *American Historical Review* 98 (October 1993): 1050–1078.

"How the Lowcountry Was Taken to Task: Slave-Labor Organization in Coastal South Carolina and Georgia." In *Slavery, Secession, and Southern History*. Robert Louis Paquette and Louis Ferleger, eds., 59–78. Charlottesville: University Press of Virginia, 2000.

"Introduction." In *Twilight on the South Carolina Rice Fields: Letters from the Heyward Family, 1862–1871*. Margaret Belser Hollis and Allen H. Stokes, eds., xvii–xxxi. Columbia: University of South Carolina Press, 2010.

Coclanis, Peter A., and J.C. Marlow. "Inland Rice Production in the South Atlantic States: A Picture in Black and White." African Americans in Southern Agriculture: 1877–1945. *Agricultural History* 72 (Spring 1998): 197–213.

Coclanis, Peter A., and Lacy K. Ford. "The South Carolina Economy Reconstructed and Reconsidered: Structure, Output, and Performance, 1670–1985." In *Developing Dixie: Modernization in a Traditional Society*. Winfred B. Moore, Jr., Joseph F. Tripp, and Lyon G. Tyler, Jr., eds., 93–110. New York: Greenwood Press, 1988.

Cohen, Hennig. "'Old Field School,' 'Cornfield School,' and 'Indian Old Field.'" *American Speech* 29 (October 1954): 226–227.

Collins, Steven G. "System, Organization, and Agricultural Reform in the Antebellum South, 1840–1860." *Agricultural History* 75 (Winter 2001): 1–27.

Colquhoun, Donald J. *Terrace Sediment Complexes in Central South Carolina*. Columbia: University of South Carolina, 1965.

Geomorphology of the Lower Coastal Plain of South Carolina. Columbia: Division of Geology, 1969.

"Cyclic Surficial Stratigraphic Units of the Middle and Lower Coastal Plains, Central South Carolina." In *Post-Miocene Stratigraphy Central and Southern*

Atlantic Coastal Plain. Robert Q. Oaks and Jules R. DuBar, eds., 179–190. Logan: Utah State University Press, 1974.

Colquhoun, Donald J., and Mark L. Brooks. "New Evidence from the Southeastern U.S. for Eustatic Components in the Late Holocine Sea Levels." *Geoarchaeology* 1 (April 1986): 275–291.

Courtenay, William A. *The Centennial of Incorporation*. Charleston: News and Courier Press, 1884.

Cowdrey, Albert E. *This Land, This South: An Environmental History, Revised Edition*. Lexington: University Press of Kentucky, 1996.

Cronon, William. *Changes in the Land: Indians, Colonists, and the Ecology of New England*. New York: Hill and Wang, 1983.

Nature's Metropolis: Chicago and the Great West. New York: W.W. Norton & Co., 1991.

"The Trouble with Wilderness; or, Getting Back to the Wrong Nature." In *Uncommon Ground: Rethinking the Human Place in Nature*. William Cronon, ed. New York: W.W. Norton & Co., 1995.

Crosby, Alfred W., Jr. *The Columbian Exchange: Biological and Cultural Consequences of 1492*. Contributions in American Studies, 2. Robert H. Walker, ed. Westport, CT: Greenwood Press, 1972.

Cross, J. Russell. *Historic Ramblin's through Berkeley County*. Columbia: R.L. Bryan Co., 1985.

Cupp, Ruth W. *Attorneys: From Charles Town to Charleston*. Birmingham, AL: Association Publishing, 2006.

Deas, Anne Simons. *Recollections of the Ball Family of South Carolina and the Comingtee Plantation*. Summerville, SC: Alwyn Ball, Jr., 1909.

Defant, Albert. *Ebb and Flow: The Tides of Earth, Air, and Water*. Ann Arbor: University of Michigan Press, 1958.

Dethloff, Henry C. "The Colonial Rice Trade." *Agricultural History* 56 (January 1982): 231– 243.

A History of the American Rice Industry, 1685–1985. College Station: Texas A&M University Press, 1988.

Doar, David. *Rice and Rice Planting in the South Carolina Lowcountry*. Charleston: Charleston Museum, 1936; reprint, 1970.

Doar, William Richardson, III, and Christopher George St. Clement Kendall. "An Analysis and Comparision of Observed Pleistocene South Carolina (USA) Shoreline Elevations with Predicted Elevations Derived from Marine Oxygen Isotope Stages." *Quaternary Research* 82 (July 2014): 164–174.

Donahue, Brian. "'Dammed at Both Ends and Cursed in the Middle': The 'Flowage' of the Concord River Meadows, 1798–1862." In *Out of the Woods: Essays in Environmental History*. Char Miller and Hal Rothman, eds., 227–242. Pittsburg: University of Pittsburg Press, 1997.

The Great Meadow: Farmers and the Land in Colonial Concord. New Haven: Yale University Press, 2004.

Downey, Tom. "Riparian Rights and Manufacturing in Antebellum South Carolina: William Gregg and the Origins of the 'Industrial Mind.'" *Journal of Southern History* 65 (February 1999): 77–108.

Duff, Meaghan N. "Creating a Plantation Province: Proprietary Land Policies and Early Settlement Patterns." In *Money, Trade, and Power: The Evolution of Colonial South Carolina's Plantation Society*. Jack P. Greene, Rosemary Brana-Shute, and Randy J. Sparks, eds., 1–25. Columbia: University of South Carolina Press, 2001.

Dunbar, Gary S. "A Southern Geographical Word List." *American Speech* 36 (December 1961): 293–296.

Dunn, Richard S. *Sugar and Slaves: The Rise of the Planter Class in the English West Indies, 1624–1713*. Chapel Hill: University of North Carolina Press, 1972.

Dusinberre, William. *Them Dark Days: Slavery in the American Rice Swamps*. New York: Oxford University Press, 1996.

Earle, Carville. "The Myth of the Southern Soil Miner: Macrohistory, Agricultural Innovation, and Environmental Change." In *The Ends of the Earth: Perspectives on Modern Environmental History*. Donald Worster, ed., 175–210. New York: Cambridge University Press, 1988.

The American Way: A Geographical History of Crisis and Recovery. Lanham, MD: Rowman and Littlefield, 2003.

Eaton, Clement. *The Mind of the Old South*. Baton Rouge: Louisiana State University Press, 1964.

Edelson, S. Max. "The Nature of Slavery: Environmental Disorder and Slave Agency in Colonial South Carolina." In *Cultures and Identities in Colonial British America*. Robert Olwell and Alan Tully, eds., 21–44. Baltimore: Johns Hopkins University Press, 2006.

Plantation Enterprise in Colonial South Carolina. Cambridge, MA: Harvard University Press, 2006.

"Clearing Swamps, Harvesting Forests: Trees and the Making of a Plantation Landscape in the Colonial South Carolina Lowcountry." *Agricultural History* 81 (Summer 2007): 381–406.

"Beyond 'Black Rice': Reconstructing Material and Cultural Contexts for Early Plantation Agriculture." *American Historical Review* 115 (February 2010): 125–135.

Edgar, Walter. *South Carolina: A History*. Columbia: University of South Carolina Press, 1998.

Eltis, David, Philip Morgan, and David Richardson. "Agency and the Diaspora in Atlantic History: Reassessing the African Contribution to Rice Cultivation in the Americas." *American Historical Review* 112 (December 2007): 1329–1358.

Evans, Chris. "The Plantation Hoe: The Rise and Fall of an Atlantic Commodity, 1650–1850." *William and Mary Quarterly* 69 (January 2012): 71–100.

Faust, Drew Gilpin. "Rhetoric and Ritual and Agriculture in Antebellum South Carolina." *Journal of Southern History* 45 (November 1979): 541–568.

James Henry Hammond and the Old South: A Design for Mastery. Baton Rouge: Lousiana State University Press, 1982.

Feeser, Andrea. *Red, White, and Black Make Blue: Indigo in the Fabric of Colonial South Carolina Life*. Athens: University of Georgia Press, 2013.

Ferguson, Leland. *Uncommon Ground: Archaeology and Early African America, 1650–1800*. Washington, D.C.: Smithsonian Institution Press, 1992.

Ferguson, Leland, and David Babson. *Survey of Plantation Sites along the East Branch of the Cooper River: A Model for Predicting Archaeological Site Location*. Columbia: University of South Carolina, 1986.

Fetters, Thomas. *Logging Railroads of South Carolina*. Forest Park, IL: Heimburger House Publishing Co., 1990.

Fiege, Mark. *Irrigated Eden: The Making of an Agricultural Landscape in the American West*. Seattle: University of Washington Press, 1999.

Fields-Black, Edda L. *Deep Roots: Rice Farmers in West Africa and the African Diaspora*. Bloomington: Indiana University Press, 2008.

"Untangling the Many Roots of West African Mangrove Rice Farming: Rice Technology in the Rio Nunez Region, Earliest Times to *c.* 1800." *Journal of African History* 49 (March 2008): 1–21.

"Atlantic Rice and Rice Farmers: Rising from Debate, Engaging New Sources, Methods, and Modes of Inquiry, and Asking New Questions." *Atlantic Studies* 12 (September 2015): 276–295.

Fogel, Robert William, and Stanley L. Engerman. *Time on the Cross: The Economics of American Negro Slavery*. 2 vols. Boston: Little, Brown, 1974.

Ford, Lacy K. "Self-Sufficiency, Cotton, and Economic Development in the South Carolina Upcountry, 1800–1860." *Journal of Economic History* 45 (June 1985): 261–267.

Fraser, Walter J., Jr. *Charleston! Charleston!: The History of a Southern City*. Columbia: University of South Carolina Press, 1989.

Lowcountry Hurricanes: Three Centuries of Storms at Sea and Ashore. Athens: University of Georgia Press, 2006.

Genovese, Eugene D. *Roll, Jordan, Roll: The World the Slaves Made*. New York: Pantheon Books, 1974.

Goldin, Claudia Dale. *Urban Slavery in the American South, 1820–1860: A Quantitative History*. Chicago: University of Chicago Press, 1976.

Goloboy, Jennifer L. "Strangers in the South: Charleston's Merchants and Middle-Class Values in the Early Republic." In *The Southern Middle Class in the Long Nineteenth Century*. Jonathan Daniel Wells and Jennifer R. Green, eds., 40–61. Baton Rouge: Louisiana State University Press, 2011.

Charleston and the Emergence of Middle-Class Culture in the Revolutionary Era. Athens: University of Georgia Press, 2016.

Graham, Otis L. "Again the Backward Region? Environmental History in and of the American South." *Southern Cultures* (Summer 2000): 50–72.

Gray, Lewis Cecil. *History of Agriculture in the Southern United States to 1860*. 2 vols. Washington, D.C.: Carnegie Institution of Washington, 1933; reprint Gloucester, MA: Peter Smith, 1958.

Green, Jennifer R. "Born into Aristocracy?: Professionals with Planter and Middle-Class Origins in Late Antebellum South Carolina." In *The Southern Middle Class in the Long Nineteenth Century*. Jonathan Daniel Wells and Jennifer R. Green, eds., 157–179. Baton Rouge: Louisiana State University Press, 2011.

Greene, Jack P. "Early South Carolina and the Psychology of British Colonization." In *Imperatives, Behaviors, and Identities: Essays in Early American Cultural History*. 87–112. Charlottesville: University of Virginia Press, 1992.

Guerard, Edward P., Jr. "The Riparian Rights Doctrine in South Carolina." *South Carolina Law Review* 21 (1969): 757–770.

Guth, Joseph H. "Law for the Ecological Age." *Vermont Journal of Environmental Law* 9 (2008): 431–512.

Gutman, Herbert G. *The Black Family in Slavery and Freedom, 1750–1925*. New York: Pantheon Books, 1976.

Haan, Richard L. "The 'Trade Do's Not Flourish as Formerly': Ecological Origins of the Yamassee War of 1715." *Ethnohistory* 28 (Autumn 1981): 341–358.

Hahn, Stephen. *Nation Under Our Feet: Black Political Struggles in the Rural South from Slavery to the Great Migration*. New York: Belknap Press, 2003.

Hall, Gwendolyn Midlo. "Africa and Africans in the African Diaspora: The Uses of Relational Databases." *American Historical Review* 115 (February 2010): 136–150.

Hardy, Stephen G. "Colonial South Carolina's Rice Industry and the Atlantic Economy: Patterns of Trade, Shipping, and Growth, 1715–1775." In *Money, Trade, and Power: The Evolution of Colonial South Carolina's Plantation Society*. Jack P. Greene, Rosemary Brana-Shute, and Randy J. Sparks, eds., 108–140. Columbia: University of South Carolina Press, 2001.

Hart, Emma. *Building Charleston: Town and Society in the Eighteenth-Century British Atlantic World*. Charlottesville: University of Virginia Press, 2010.

Hart, John F. "Colonial Land Use Law and Its Significance for Modern Takings Doctrine." *Harvard Law Review* 109 (April 1996): 1252–1300.

Hawley, Norman R. "The Old Rice Plantations in and around the Santee Experimental Forest." *Agricultural History* 23 (April 1949): 86–91.

Hawthorne, Walter. "From 'Black Rice' to 'Brown': Rethinking the History of Risiculture in the Seventeenth and Eighteenth Century Atlantic." *American Historical Review* 115 (February 2010): 151–163.

Haygood, Tamara Miner. *Henry William Ravenel, 1814–1887: South Carolina Scientist in the Civil War Era*. Tuscaloosa: University of Alabama Press, 1987.

Hays, Samuel P. *Explorations in Environmental History*. Pittsburgh: University of Pittsburgh Press, 1998.

Heitzler, Michael J. *Goose Creek: A Definitive History*. Volume I: *Planters, Politicians, and Patriots*. Charleston: History Press, 2005.

Hester, Al. "Establishing the Francis Marion: National Forest History in South Carolina's Lowcountry, 1901–1936." *Forest History Today* 17 (Spring/Fall 2011): 56–63.

Heyward, Duncan Clinch. *Seed from Madagascar*. Chapel Hill: University of North Carolina Press, 1937; reprint, Columbia: University of South Carolina Press, 1993.

Hilliard, Sam B. "The Tidewater Rice Plantation: An Ingenious Adaptation to Nature." In *Geoscience and Man*, vol. 12. H.J. Walker, ed., 57–66. Baton Rouge: Louisiana State Press, 1975.

"Antebellum Tidewater Rice Culture in South Carolina and Georgia." In *European Settlement and Development in North America: Essays on Geographical Change in Honour and Memory of Andrew Hill Clark*. James R. Gibson, ed., 91–115. Toronto: University of Toronto Press, 1978.

Hirsch, Arthur H. *The Huguenots of South Carolina*. Durham: Duke University Press, 1928; reprint, Columbia: University of South Carolina Press, 1999.

Hodges, John D. "Minor Alluvial Floodplains." In *Southern Forested Wetlands: Ecology and Management*. Michael G. Messina and William H. Conner, eds., 325–341. Boca Raton, FL: Lewis Publishers, 1998.

Horwitz, Morton J. *The Transformation of American Law, 1780–1860*. Cambridge, MA: Harvard University Press, 1977.

House, Albert V., Jr. "Charles Manigault's Essay on the Open Planting of Rice." *Agricultural History* 16 (October 1942): 184–193.

Planter Management and Capitalism in Ante-bellum Georgia. New York: Columbia University Press, 1954.

Irving, Aemilius, and L. Homfray, ed. *James Irving of Ironshore and His Descendants, 1713–1918*. Toronto, ON: College Press Ltd, 1918.

Johnny, Michael, John Karimu, and Paul Richards. "Upland Swamp Rice Farming Systems in Sierra Leone: The Social Context of Technological Change." *Africa: Journal of the International African Institute* 51 (April 1981): 596–620.

Jones, Alice Hanson. *Wealth of a Nation to Be: The American Colonies on the Eve of the Revolution*. New York: Columbia University Press, 1980.

Jones-Jackson, Patricia. *When Roots Die: Endangered Traditions on the Sea Islands*. Athens: University of Georgia Press, 1987.

Joseph, J.W. "Building to Grow: Agrarian Adaptations to South Carolina's Historical Landscapes." In *Carolina's Historical Landscapes: Archaeological Perspectives*. Linda F. Stine et al., eds., 45–60. Knoxville: University of Tennessee Press, 1997.

Joyner, Charles. *Down by the Riverside: A South Carolina Slave Community*. Urbana: University of Illinois Press, 1984.

Kapsch, Robert J. *Historic Canals and Waterways of South Carolina*. Columbia: University of South Carolina Press, 2010.

Kerridge, Eric. *The Agricultural Revolution*. London: Allen & Unwin, 1967.

Kirby, Jack Temple. *Poquosin: A Study of Rural Landscape and Society*. Chapel Hill: University of North Carolina Press, 1995.

Mockingbird Song: Ecological Landscapes of the South. Chapel Hill: University of North Carolina Press, 2006.

Knight, Frederick C. *Working the Diaspora: The Impact of African Labor on the Anglo-American World, 1650–1850*. New York: New York University Press, 2010.

Koger, Larry. *Black Slaveowners: Free Blacks Masters in South Carolina, 1790–1860*. Jefferson, NC: McFarland & Co, 1985.

Kolchin, Peter. "The World the Historians Made: Peter Wood's Black Majority in Historiographical Context." *South Carolina Historical Magazine* 100 (October 1999): 368–378.

Kovacik, Charles F. "Land Use Change on the South Carolina Rice Plantation Lands." In *Historical Geography '76: International Geographical Congress, 23rd Moscow 1976*, 64–67. Oxford: Pergamon Press, 1976.

"South Carolina Rice Coast Landscape Changes." In *Proceedings: Tall Timbers and Management Conference, Number 16, February 22–24, 1979, Thomasville, GA*. Tallahassee, FL: Tall Timbers Research Station, 1979.

Kovacik, Charles F., and John J. Winberry. *South Carolina: The Making of a Landscape*. Boulder: Westview Press, 1987; reprint, Columbia: University of South Carolina Press, 1989.

Kulik, Gary. "Dams, Fish, and Farmers: Defense of Public Rights in Eighteenth Century Rhode Island." In *The Countryside in the Age of Capitalist Transformation*. Stephen Hahn and Jonathan Purdue, eds., 25–50. Chapel Hill: University of North Carolina Press, 1985.

Kulikoff, Allan. *Tobacco and Slaves: The Development of Southern Cultures in the Chesapeake, 1680–1800*. Chapel Hill: University of North Carolina Press, 1986.

Lauer, T.E. "Common Law Background of the Riparian Doctrine." *Missouri Law Review* 28 (1963): 60–107.

Lees, William. *Limerick: Old and in the Way, Archeological Investigations at Limerick Plantation, Berkley County, South Carolina*. Columbia: Institute of Archaeology and Anthropology, 1980.

Leiding, Harriette Kershaw. *Historic Houses of South Carolina*. Philadelphia: J.B. Lippincott, 1921.

Leopold, Aldo. *A Sand County Almanac*. New York: Oxford University Press, 1949.

Linares, Olga F. "From Tidal Swamp to Inland Valley: On the Social Organization of Wet Cultivation among the Diola of Senegal." *Africa: Journal of the International African Institute* 51 (April 1981): 557–595.

"African Rice (*Oryza glaberrima*): History and Future Potential." *Proceedings of the National Academy of Sciences* 99 (2002): 16360–16365.

Linder, Suzanne Cameron. *Historical Atlas of the Rice Plantations of the ACE River Basin*. Columbia: South Carolina Department of Archives and History, 1995.

Anglican Churches in Colonial South Carolina: Their History and Architecture. Charleston: Wyrick & Co., 2000.

Littlefield, Daniel C. *Rice and Slaves: Ethnicity and the Slave Trade in Colonial South Carolina*. Baton Rouge: Louisiana State University Press, 1981.

Rice and the Making of South Carolina: An Introductory Essay. Columbia: South Carolina Department of Archives and History, 1995.

Majewski, John D. *Modernizing a Slave Economy: The Economic Vision of the Confederate Nation*. Chapel Hill: University of North Carolina Press, 2009.

Mancall, Peter C., et al. "Agricultural Labor Productivity in the Lower South, 1720–1800." *Explorations in Economic History* 39 (October 2002): 390–424.

Marshall, Bill. *Ashley Scenic River Management Plan*. Columbia: South Carolina Department of Natural Resources, 2003.

Martin, Calvin Luther. *In the Spirit of the Earth: Rethinking History and Time.* Baltimore: Johns Hopkins University Press, 1992.

Martins, Susanna Wade, and Tom Williamson. "Floated Water-Meadows in Norfolk: A Misplaced Innovation." *Agricultural History Review* 42 (1994): 20–37.

Mathew, William M. *Edmund Ruffin and the Crisis of Slavery in the Old South: The Failure of Agricultural Reform.* Athens: University of Georgia Press, 1988.

McCandless, Peter. *Slavery, Disease, and Suffering in the Southern Lowcountry.* New York: Cambridge University Press, 2011.

McClain, Molly, and Alessa Ellefson. "A Letter from Carolina, 1688: French Huguenots in the New World." *William and Mary Quarterly* 64 (April 2007): 377–394.

McCrady, Edward. *The History of South Carolina under the Royal Government, 1719–1776.* London: Macmillan, 1899.

McCusker, John J. "Table Eg1160–1165 Rice exported from South Carolina and Georgia: 1698–1790." In *Historical Statistics of the United States*, Millennial Edition Online. Susan B. Carter, Scott Sigmund Gartner, Michael R. Haines, Alan L. Olmstead, Richard Sutch, and Gavin Wright, eds. New York: Cambridge University Press, 2006.

Meanly, Brooke. *Swamps, River Bottoms, and Canebrakes.* Barre, MA: Barre Publishers, 1972.

Meinig, D.W. *The Shaping of America: A Geological Perspective on 500 Years of History.* Volume I: *Atlantic America, 1492–1800.* New Haven: Yale University Press, 1986.

The Shaping of America: A Geological Perspective on 500 Years of History. Volume II: *Continental America, 1800–1867.* New Haven: Yale University Press, 1993.

Menard, Russell R. "Slavery, Economic Growth, and Revolutionary Ideology in the South Carolina Lowcountry." In *The Economy of Early North America: The Revolutionary Period, 1763–1790.* Ronald Hoffman, John J. McCusker, Russell R. Menard, and Peter J. Albert, eds., 244–274. Charlottesville: University Press of Virginia, 1988.

"Financing the Lowcountry Export Boom: Capital and Growth in Early South Carolina." *William and Mary Quarterly* 51 (October 1994): 659–676.

"Economic and Social Development of the South." In *The Cambridge Economic History of the United States.* Volume I: *The Colonial Era.* Stanley L. Engerman and Robert E. Gallman, eds., 249–295. Cambridge, UK: Cambridge University Press, 1996.

Merchant, Carolyn. "Reinventing Eden: Western Culture as a Recovery Narrative." In *Uncommon Ground: Rethinking the Human Place in Nature.* William Cronon, ed., 132–159. New York: W.W. Norton & Co., 1995.

Meriwether, Robert L. *The Expansion of South Carolina: 1729–1765.* Kingsport, TN: Southern Publishers, 1940.

Merrens, H. Roy, and Gerorge D. Terry. "Dying in Paradise: Malaria, Mortality, and the Perceptual Environment in Colonial South Carolina." *Journal of Southern History* 50 (November 1984): 533–550.

"Middleton." *South Carolina Historical and Genealogical Magazine* 1 (July 1900): 228–261.

Miller, Char. *Gifford Pinchot and the Making of Modern Environmentalism*. Washington, D.C.: Island Press, 2001.

Miller, James David. *South by Southwest: Planter Emigration and Identity in the Slave South*. Charlottesville: University of Virginia Press, 2002.

Mock, Cary J. "Documentary Records of Past Climate and Tropical Cyclones from the Southeastern United Sates," *PAGES News* 10 (December 2002): 21–22.

"Tropical Cyclone Reconstructions from Documentary Records: Examples for South Carolina, United States." In *Hurricanes and Typhoons: Past, Present, and Future*. Richard J. Murnane and Kam-Biu Liu, eds., 121–176. New York: Columbia University Press, 2004.

Moore, Alexander. "Daniel Axtell's Account Book and the Economy of Early South Carolina," *South Carolina Historical Magazine* 95 (October 1994): 280–301.

Morgan, Edmund S. *American Slavery, American Freedom: The Ordeal of Colonial Virginia*. New York: W. W. Norton & Co., 1975.

Morgan, Philip D. "A Profile of a Mid-Eighteenth Century South Carolina Parish: The Tax Return of St. James', Goose Creek." *South Carolina Historical Magazine* 81 (January 1980): 51–65.

"Work and Culture: The Task System and the World of Lowcountry Blacks, 1700 to 1880." *William and Mary Quarterly* 39 (October 1982): 563–599.

Slave Counterpoint: Black Culture in the Eighteenth-Century Chesapeake and Lowcountry. Chapel Hill: University of North Carolina Press, 1998.

Morris, Christopher. "A More Southern Environmental History." *Journal of Southern History* 75 (August 2009): 581–598.

Mouser, Bruce, Edwin Nuijten, and Paul Richards. "A Colour Bar? The Commercialization of Rice at Sierra Leone." New Histories of Rice, Working Paper. Max Planck Institute for the History of Science, Berlin, Germany, 25–26 March 2011.

Mullin, Gerald W. *Flight and Rebellion: Slave Resistance in Eighteenth-Century Virginia*. New York: Oxford University Press, 1972.

Mullin, Michael, ed. *American Negro Slavery: A Documentary History*. Columbia: University of South Carolina Press, 1976.

Nadelshaft, Jerome J. *The Disorders of War: The Revolution in South Carolina*. Orono: University of Maine Press, 1981.

Nash, R.C. "South Carolina and the Atlantic Economy in the Late Seventeenth and Eighteenth Centuries." *Economic History Review*, New Series 45 (November 1992): 677–702.

"The Organization of Trade and Finance in the Atlantic Economy: Britain and South Carolina, 1670–1775." In *Money, Trade, and Power: The Evolution of Colonial South Carolina's Plantation Society*. Jack P. Greene, Rosemary Brana-Shute, and Randy J. Sparks, eds., 74–107. Columbia: University of South Carolina Press, 2001.

Nash, Roderick. *Wilderness and the American Mind*. 3rd ed. New Haven: Yale University Press, 1982.

Nelson, Lynn A. *Pharsalia: An Environmental Biography of a Southern Plantation, 1780–1880*. Athens: University of Georgia Press, 2007.

Nelson, William E. *Americanization of Common Law: The Impact of Legal Change on Massachusetts Society, 1760–1830*. Cambridge, MA: Harvard University Press, 1975.

Oakes, James. *The Ruling Race: A History of American Slaveholders*. New York: Vantage Books, 1982.

Oates, Stephen B. *The Fires of Jubilee: Nat Turner's Fierce Rebellion*. New York: Harper & Row, 1975.

Olwell, Robert. *Masters, Slaves, and Subjects: The Culture of Power in the South Carolina Lowcountry, 1740–1790*. Ithaca: Cornell University Press, 1998.

O'Malley, Gregory E. "Diversity in the Slave Trade to the Colonial Carolinas." In *Creating and Contesting Carolina: Proprietary Era Histories*. Michelle LeMaster and Bradford J. Wood, eds., 234–255. Columbia: University of South Carolina Press, 2013.

O'Neall, John Belton. *Biographical Sketches of the Bench and Bar of South Carolina*. Volume 1. Charleston: S.G. Courtenay & Co., 1859.

Opie, John. *Nature's Nation: An Environmental History of the United States*. Fort Worth: Harcourt Brace, 1998.

Otto, John S. "The Origins of Cattle-Ranching in Colonial South Carolina, 1670–1715." *South Carolina Historical Magazine* 87 (April 1986): 117–124.

"Livestock-Raising in Early South Carolina, 1670–1700: Prelude to the Early Rice Plantation Economy." *Agricultural History* 61 (1987): 13–24.

The Southern Frontiers, 1607–1860: The Agricultural Evolution of the Colonial and Antebellum South. New York: Greenwood Press, 1989.

Pennington, Patience. *Woman Rice Planter*. New York: Macmillan Company, 1914.

Pfister, Christian. "Climate and Economy in Eighteenth-Century Switzerland." *Journal of Interdisciplinary History* 9 (Autumn 1978): 223–243.

Phillips, Ulrich Bonnell. *Life and Labor in the Old South*. Boston: Little, Brown, and Company, 1929.

Pinckney, Elise. "Still Mindful of the English Way: 250 Years of Middleton Place on the Ashley." *South Carolina Historical Magazine* 92 (July 1991): 149–171.

Pisani, Donald J. "Beyond the Hundredth Meridian: Nationalizing the History of Water in the United States." *Environmental History* 5 (October 2000): 466–482.

Porcher, F.A. *The History of the Santee Canal*. Charleston: South Carolina Historical Society, 1875; reprint, 1903.

Porcher, Richard D. *A Teacher's Field Guide to the Natural History of the Bluff Plantation Wildlife Sanctuary*. New Orleans: Kathleen O'Brien Foundation, 1985.

"Rice Culture in South Carolina: A Brief History, the Role of the Huguenots, and the Preservation of Its Legacy." *Transactions of the Huguenot Society of South Carolina* 92 (1987): 1–22.

Porcher, Richard D., and Douglas A. Rayner. *A Guide to the Wildflowers of South Carolina*. Columbia: University of South Carolina Press, 2001.

Porcher, Richard D., and Sarah Fick. *The Story of Sea Island Cotton*. Charleston: Wyrick Press, 2005.

Porcher, Richard D., and William Robert Judd. *The Market Preparation of Carolina Rice*. Columbia: University of South Carolina Press, 2014.

Postell, William Dosite. *The Health of Slaves on Southern Plantations*. Baton Rouge: Louisiana State University Press, 1951; reprint, Peter Smith: Gloucester, MA, 1970.

Price, Richard. "Subsistence on the Plantation Periphery: Crops, Cooking, and Labour among Eighteenth-Century Suriname Maroons." *Slavery and Abolition* 12 (1, 1991): 107–127.

Pringle, Elizabeth W. Allston. *Chronicles of Chicora Wood*. New York: Charles Scribner's Sons, 1922.

Pyne, Stephen. *Vestal Fire: An Environmental History, Told through Fire, of Europe and Europe's Encounter with the World*. Seattle: University of Washington Press, 1997.

Ramsay, David. *Ramsay's History of South Carolina: From Its First Settlement in 1670 to the Year 1808*. 2 vols. Charleston: Walker, Evans, and Co., 1858.

Ravenel, Daniel. "Historical Sketch of Huguenot Congregations of South Carolina." *Transactions of the French Huguenot Society* 7 (1900): 7–74.

Rawick, George P. *From Sundown to Sunup: The Making of the Black Community*. Westport: Greenwood Publishing Co., 1972.

Recknagel, A.B. "Remeasuring the Hell Hole Plots in South Carolina." *Journal of Forestry* 26 (1928): 823–824.

Rehder, John B. *Delta Sugar: Louisiana's Vanishing Plantation Landscape*. Baltimore: Johns Hopkins University Press, 1999.

Reithmaier, Tina. "Maps and Photographs." In *The Historical Ecology Handbook: A Restorationist's Guide to Reference Ecosystems*. Dave Egan and Evelyn A. Howell, eds. Washington D.C.: Island Press, 2001.

Rodgers, George C., Jr. *History of Georgetown County, South Carolina*. Columbia: University of South Carolina Press, 1970.

Roper, L.H. *Conceiving Carolina: Proprietors, Planters, and Plots, 1662–1729*. New York: Palgrave Macmillan, 2004.

Rosengarten, Dale. *Row upon Row: Sea Grass Baskets of the South Carolina Lowcountry*. Columbia: McKissick Museum, University of South Carolina, 1986.

Rosengarten, Dale, and Theodore Rosengarten, eds. *Portion of the People: Three Hundred Years of Southern Jewish Life*. Columbia: University of South Carolina Press, 2002.

Rosengarten, Theodore. "*The Southern Agriculturalist* in the Age of Reform." In *Intellectual Life in Antebellum Charleston*. Michael O'Brien and David Moltke-Hansen, eds., 279–294. Knoxville: The University of Tennessee Press, 1986.

Tombee: Portrait of a Cotton Planter. New York: William Marrow, 1986.

"In the Master's Garden." In *Art and Landscape in Charleston and the Lowcountry*. John Beardslay, ed. Washington D.C.: Spacemaker Press, 1998.

Rowland, Lawrence S., Alexander Moore, and George C. Rodgers, Jr. *The History of Beaufort County, South Carolina*. Volume I: *1514–1861*. Columbia: University of South Carolina Press, 1996.

Russell, Emily W. B. *People and the Land through Time: Linking Ecology and History*. New Haven: Yale University Press, 1997.

Ryden, David B., and Russell R. Menard. "South Carolina's Colonial Land Market: An Analysis of Rural Property Sales, 1720–1775." *Social Science History* 29 (Winter 2005): 599–623.

Salley, Alexander Samuel Jr. *The Introduction of Rice Culture in South Carolina*. Historical Commission of South Carolina, Bulletin 6. Columbia: State Company, 1919.

Sanders, Albert E., and William D. Anderson, Jr. *Natural History Investigations in South Carolina: From Colonial Times to the Present*. Columbia: University of South Carolina Press, 1999.

Sass, Herbert Ravenel. "The Low-Country Lagoons." *Charleston Museum Quarterly* 2 (First Quarter 1932): 20–23.

Savage, Henry, Jr. "Carolina Golden Rice." In *Perspectives in South Carolina History: The First 300 Years*. Ernest M. Lander, Jr. and Robert K. Ackerman, eds. Columbia: University of South Carolina Press, 1973.

Scarborough, William Kauffman. "Science on the Plantation." In *Science and Medicine in the Old South*. Ronald L. Numbers and Todd L. Savitt, eds., 79–106. Baton Rouge: Louisiana State Press, 1989.

Masters of the Big House: Elite Slaveholders of the Mid-Nineteenth-Century South. Baton Rouge: Louisiana State Press, 2003.

Schantz, Mark S. "'A Very Serious Business': Managerial Relationships on the Ball Plantations, 1800–1835." *South Carolina Historical Magazine* 88 (January 1987): 1–22.

Schwalm, Leslie A. *A Hard Fight for We: Women's Transition from Slavery to Freedom in South Carolina*. Urbana and Chicago: University of Illinois Press, 1997.

Shanks, Judith Alexander Weil. *Old Family Things: An Affectionate Look Back*. www.judithwshanks.com/old_family_things/index.html

Shatzman, Aaron M. *Servants into Planters: The Origin of an American Image; Land Acquisition and Status Mobility in 17th-Century South Carolina*. New York: Garland Publishers, 1989.

Shuler, Jack. *Calling Out Liberty: The Stono Slave Rebellion and the Universal Struggle for Human Rights*. Oxford: University of Mississippi Press, 2009.

Silver, Timothy. *A New Face on the Countryside: Indians, Colonists, and Slaves in South Atlantic Forests, 1500–1800*. Cambridge, UK: Cambridge University Press, 1990.

Simms, William Gilmore. *The Geography of South Carolina: Being a Companion to the History of the State*. Charleston: Babcock & Co., 1843.

Singleton, Theresa A., ed. *The Archaeology of Slavery and Plantation Life*. Orlando: Academic Press, 1985.

"Reclaiming the Gullah-Geechee Past: Archaeology of Slavery in Coastal Georgia." In *African American Life in the Georgia Lowcountry: The Atlantic*

World and the Gullah Geechee. Philip Morgan, ed., 151–187. Athens: University of Georgia Press, 2010.

Siple, George E. "Some Geologic and Hydrologic Factors Affecting Limestone Terraces of Tertiary Age in South Carolina." *Southeastern Geology* 2 (August 1960): 1–11.

Sirmans, M. Eugene. *Colonial South Carolina: A Political History, 1663–1763*. Chapel Hill: University of North Carolina Press, 1966.

Sluyter, Andrew. *Black Ranching Frontiers: African Cattle Herders of the Atlantic World, 1500–1900*. New Haven: Yale University Press, 2012.

Smith, Alfred Glaze. *Economic Readjustment of an Old Cotton State: South Carolina, 1820–1860*. Columbia: University of South Carolina Press, 1958.

Smith, Henry A.M. "The Baronies of South Carolina: Seewee Barony." *South Carolina Historical and Genealogical Magazine* 12 (July 1911): 109–117.

"The Baronies of South Carolina: Landgrave Ketelby's Barony." *South Carolina Historical and Genealogical Magazine* 15 (October 1914): 149–165.

"The Baronies of South Carolina: Quimby and the Eastern Branch of Cooper River." *South Carolina Historical and Genealogical Magazine* 18 (January 1917): 3–36.

"Charleston and Charleston Neck." *South Carolina Historical and Genealogical Magazine* 19 (January 1918): 3–76.

"The Ashley River: Its Seats and Settlements." *South Carolina Historical and Genealogical Magazine* 20 (April 1919): 75–122.

"The Upper Ashley: And the Mutations of Families." *South Carolina Historical and Geological Magazine* 20 (July 1919): 151–198.

Smith, Julia F. *Slavery and Rice Culture in Low Country Georgia, 1750–1860*. Knoxville: University of Tennessee Press, 1985.

Smith, Mark M. "Time, Slavery and Plantation Capitalism in the Ante-bellum American South." *Past and Present* 150 (February 1996): 142–169.

Mastered by the Clock: Time, Slavery, and Freedom in the American South. Chapel Hill: University of North Carolina Press, 1997.

Snyder, Terri L. "Legal History of the Colonial South: Assessment and Suggestions." *William and Mary Quarterly* 50 (January 1993): 18–27.

Soller, David R., and Hugh H. Mills. "Surficial Geology and Geomorphology." In *The Geology of the Carolinas: Carolina Geological Society Fiftieth Anniversary Volume*. J.W. Horton and V.A. Zullo, eds., 290–308. Knoxville: University of Tennessee Press, 1991.

South, Stanley, and Michael Hartley. *Deep Water and High Ground: Seventeenth Century Low Country Settlement*. Research Manuscript Series 166. Columbia: University of South Carolina Institute of Archaeology and Anthropology, 1980.

Sprunt, Alexander, Jr., and E. Burnham Chamberlain. *South Carolina Bird Life*. Rev. ed. Columbia: University of South Carolina Press, 1970.

Steffen, Charles G. "In Search of the Good Overseer: The Failure of the Agricultural Reform Movement in Lowcountry South Carolina, 1821–1834." *Journal of Southern History* 63 (November 1997): 753–802.

Steinberg, Ted. *Nature Incorporated: Industrialization of the Waters of New England*. New York: Cambridge University Press, 1991.

Acts of God: The Unnatural History of Natural Disaster in America. New York: Oxford University Press, 2000.

Down to Earth: Nature's Role in American History. New York: Oxford University Press, 2002.

Stephens, Lester D. *Science, Race, and Religion in the American South: John Bachman and the Charleston Circle of Naturalists, 1815–1895*. Chapel Hill: University of North Carolina Press, 2000.

Stewart, Mart A. "'Whether Wast, Deodand, or Stray': Cattle, Culture, and the Environment in Early Georgia." *Agricultural History* 65 (Summer 1991): 1–28.

"Rice, Water, and Power: Landscapes of Domination and Resistance in the Lowcountry, 1780–1880." *Environmental History Review* 15 (Fall 1991): 47–64.

"What Nature Suffers to Groe": Life, Labor, and Landscape on the Georgia Coast, 1680–1920. Athens: University of Georgia Press, 1996.

"Let Us Begin with the Weather?': Climate, Race, and Cultural Distinctiveness in the American South." In *Nature and Society in Historical Context*. Mikulas Teich, ed., 240–256. Cambridge, UK: Cambridge University Press, 1997.

"If John Muir Had Been an Agrarian: Environmental History West and South," *Environment and History* 11 (May 2005): 139–162

Stoll, Steven. *Larding the Lean Earth: Soil and Science in Nineteenth-Century America*. New York: Hill and Wang, 2002.

Sutter, Paul S. "No More the Backward Region: Southern Environmental History Comes of Age." In *Environmental History and the American South: A Reader*. Paul S. Sutter and Christopher Manganiello, eds., 1–24. Athens: University of Georgia Press, 2009.

"What Gullies Mean: Georgia's 'Little Grand Canyon' and Southern Environmental History." *Journal of Southern History* 76 (August 2010): 579–616.

Taylor, David. *South Carolina Naturalists: An Anthology, 1700–1860*. Columbia: University of South Carolina Press, 1998.

Tiner, Ralph W. *Field Guide to Coastal Wetland Plants of Southeastern United States*. Amherst: University of Massachusetts Press, 1993.

In Search of Swampland: A Wetland Sourcebook and Field Guide. New Brunswick: Rutgers University Press, 1998.

Trinkley, Michael, Natalie Adams, and Debi Hacker. *Landscape and Garden Archaeology at Crowfield Plantation: A Preliminary Investigation*, Research Series 32. Columbia: Chicora Foundation, 1992.

Tuten, James H. *Lowcountry Time and Tide: The Fall of the South Carolina Rice Kingdom*. Columbia: University of South Carolina Press, 2010.

Valencius, Conevery Bolton. *The Health of the Country: How American Settlers Understood Themselves and Their Land*. New York: Basic Books, 2002.

Van Ruymbeke, Bertrand. "Huguenots of Proprietary South Carolina: Patterns of Migration and Integration." In *Money, Trade, and Power: The Evolution of Colonial South Carolina's Plantation Society*. Jack P. Greene, Rosemary Brana-Shute, and Randy J. Sparks, eds., 26–48. Columbia: University of South Carolina Press, 2001.

From New Babylon to Eden: The Huguenots and Their Migration to Colonial South Carolina. Columbia: The University of South Carolina Press, 2006.

Vernon, Amelia Wallace. *African Americans at Mars Bluff, South Carolina*. Baton Rouge: Louisiana State University Press, 1993.

Ver Steeg, Clarence L. *Origins of a Southern Mosaic: Studies of Early Carolina and Georgia*. Athens: University of Georgia Press, 1975.

Vileisis, Ann. *Discovering the Unknown Landscape: A History of America's Wetlands*. Washington, D.C.: Island Press, 1997.

Vlach, John Michael. *Back of the Big House: The Architecture of Plantation Slavery*. Chapel Hill: University of North Carolina Press, 1993.

Waddell, Gene. *Indians in the South Carolina Lowcountry, 1562–1751*. Columbia: University of South Carolina, 1980.

Wallace, Anthony F. C. *Rockdale: The Growth of an American Village in the Early Industrial Revolution*. New York: Alfred A. Knopf, 1978.

Wallace, David Duncan. *The History of South Carolina*. 4 vols. New York: American Historical Society, 1934.

Waring, Joseph I. *A History of Medicine in South Carolina, 1620–1825*. Charleston: South Carolina Medical Association, 1964.

Watson, Harry L. "'The Common Rights of Mankind': Subsistence, Shad, and Commerce in the Early Republican South." *Journal of American History* 83 (June 1996): 13–43.

Way, Albert G. *Conserving Southern Longleaf: Herbert Stoddard and the Rise of Ecological Management*. Athens: University of Georgia Press, 2011.

Wayne, Arthur T. "The Nest and Eggs of Bachman's Warbler, *Helminthophila bachmani* (Aud.), Taken near Charleston, South Carolina." *The Auk* 24 (January 1907): 43–48.

Wayne, Lucy B. "'Burning Brick and Making a Large Fortune at It Too': Landscape Archaeology and Lowcountry Brickmaking." In *Carolina's Historical Landscapes: Archaeological Perspectives*. Linda F. Stine, Martha Zierden, Lesley M. Drucker, and Christopher Judge, eds., 97–112. Knoxville: University of Tennessee Press, 1997.

Webber, Mabel L. "Inscriptions from the Church-Yard at Strawberry Chapel." *South Carolina Historical and Genealogical Magazine* 21 (October 1920): 161–170.

Weir, Robert M. *Colonial South Carolina: A History*. New York: KTO Press, 1983; reprint, Columbia: University of South Carolina Press, 1997.

White, Richard. *Land Use, Environment, and Social Change: The Shaping of Island County, Washington*. Seattle: University of Washington Press, 1980.

The Middle Ground: Indians, Empires, and Republics in the Great Lakes Region, 1650–1815. New York: Cambridge University Press, 1991.

The Organic Machine: The Remaking of the Columbia River. New York: Hill and Wang, 1995.

Whitney, Gordon G. *From Coastal Wilderness to Fruited Plain: A History of Environmental Change in Temperate North America from 1500 to the Present*. Cambridge, UK: Cambridge University Press, 1994.

Wilms, Douglas C. "The Development of Rice Culture in Eighteenth Century Georgia." *Southeastern Geographer* 12 (January 1972): 45–57.

Wood, Betty. *Women's Work, Men's Work: The Informal Slave Economies of Lowcountry Georgia*. Athens: University of Georgia Press, 1995.

Wood, Peter H. *Black Majority: Negroes in Colonial South Carolina from 1670 through the Stono Rebellion*. New York: W. W. Norton & Co., 1974.

"They Understand Their Business Well: West Africans in Early South Carolina." In *Grass Roots: African Origins of an American Art*. Dale Rosengarten, Theodore Rosengarten, Enid Schildkrout, eds., 78–93. Seattle: University of Washington Press, 2008.

Worster, Donald. *Rivers of Empire: Water, Aridity, and the Growth of the American West*. New York: Oxford University Press, 1985.

The Wealth of Nature: Environmental History and the Ecological Imagination. New York: Oxford University Press, 1993.

Nature's Economy: A History of Ecological Ideas, 2nd ed. New York: Cambridge University Press, 1994.

Zahniser, Marvin. *Charles Cotesworth Pinckney: Founding Father*. Chapel Hill: University of North Carolina Press, 1967.

Zierden, Martha, Suzanne Linder, and Ronald Anthony. *Willtown: An Archaeological and Historical Perspective*. The Charleston Museum Archaeological Contributions 27. Columbia: South Carolina Department of Archives and History, 1999.

Zierden, Martha A., and Elizabeth J. Reitz. *Archaeology at City Hall: Charleston's Colonial Beef Market*. Archaeological Contributions 34. Charleston: The Charleston Museum, 2005.

Theses and Dissertations

Alsup, Dianna Lynn. "A Study of Plant Distribution Changes in Georgetown County Abandoned Rice Fields: Analyzing Succession in Freshwater Estuaries Using Remote Sensing and Geographic Information System (GIS) Applications." M.S. thesis, University of Charleston; Medical University of South Carolina, 1999.

Anthony, Ronald W. "Cultural Diversity at Mid to Late Eighteenth Century Lowcountry Plantation Slave Settlements." M.A. thesis, University of South Carolina, 1989.

Babson, David W. "The Tanner Road Settlement: The Archaeology of Racism on Limerick Plantation." M.A. thesis, University of South Carolina, 1987.

Bailey, Ralph, Jr. "A History of the South Carolina Lowcountry Plantation Landscape, 1663–1920: An Example of the Use of Plants and Plant Collections as Historical Resources." M.A. thesis, University of Charleston; The Citadel, 1997.

Carpenter, James G. "The Rice Plantation Lands of Georgetown County, South Carolina: A Historical Geographic Study." M.A. thesis, University of South Carolina, 1973.

Clow, Richard. "Edward Rutledge of South Carolina." Ph.D. diss., University of Georgia, 1976.

Cody, Cheryll Ann. "Slave Demography and Family Formation: A Community Study of the Ball Family Plantations, 1720–1896." Ph.D. diss., University of Minnesota, 1982.

Coon, David L. "The Development of Market Agriculture in South Carolina, 1670–1785." Ph.D. diss., University of Illinois at Urbana-Champaign, 1972.

Crouse, Maurice Alfred. "The Manigault Family of South Carolina, 1685–1783." Ph.D. diss., Northwestern University, 1964.

Doar, William Richardson, III. "The Geologic Implications of the Factors That Affected Relative Sea-Level Positions in South Carolina during the Pleistocene and Associated Preserve High-stand Deposits." Ph.D. diss., University of South Carolina, 2014.

Duff, Meaghan N. "Designing Carolina: The Construction of an Early American Social and Geographical Landscape, 1670–1719." Ph.D. diss., William & Mary, 1998.

Edelson, S. Max. "Planting the Lowcountry: Agricultural Enterprise and Economic Experience in the Lower South, 1695–1785." Ph.D. diss., Johns Hopkins University, 1998.

Groening, Richard I., Jr. "The Rice Landscape in South Carolina: Valuation, Technology, and Historical Periodization." M.A. thesis, University of South Carolina, 1998.

Hart, T. Robert. "Santee-Cooper Landscape: Culture and Environment in the South Carolina Lowcountry." Ph.D. diss., University of Alabama, 2004.

Hewitt, Gary Livingston. "Expansion and Improvement: Land, People, and Politics in South Carolina and Georgia, 1690–1745." Ph.D. diss., Princeton, 1996.

Hunt, Judith Lee. "The Circle Large and Small: Kinship, Slavery, and the Middleton Plantocracy, a Lowcountry Example." M.A. thesis, University of Charleston; The Citadel, 1995.

Lane, George Winston, Jr. "The Middletons of Eighteenth Century South Carolina: A Colonial Dynasty, 1678–1787." Ph.D. diss., Emory University, 1990.

Lee, Jeon. "The Historical Geography of Rice Culture in the American South." Ph.D. diss., Louisiana State University, 1988.

Morgan, Patrick H. "A Study of Tidelands and Improvements within a Three-River Delta System – The South Edisto, Ashepoo, and Combahee Rivers of South Carolina." M.A. thesis, University of Georgia, 1974.

Oatis, Stephen J. "A Colonial Complex: South Carolina's Changing Era of the Yamasee War, 1680–1730." Ph.D. diss., Emory University, 1999.

Pett-Conklin, Linda M. "Cadastral Surveying in Colonial South Carolina: A Historical Geography." Ph.D. diss., Louisiana State University, 1986.

Senese, Donald Joseph. "Legal Thought in South Carolina, 1800–1860." Ph.D. diss., University of South Carolina, 1970.

Shlasko, Ellen. "Carolina Gold: Economic and Social Change on a South Carolina Rice Plantation." Ph.D. diss., Yale University, 1997.

Smith, Hayden R. "Watersheds of Control: An Environmental History of South Carolina Lowcountry Rice Plantations, 1760–1860." M.A. thesis, University of Charleston, 2002.

Smith, Shelly Elizabeth. "The Plantations of Colonial South Carolina: Transmission and Transformation in Provincial Culture." Ph.D. diss., Columbia University, 1999.

Terry, George D. "'Champaign Country': A Social History of an Eighteenth Century Lowcountry Parish in South Carolina, St. Johns Berkley County." Ph.D. diss., University of South Carolina, 1981.

Willis, Russell Austin. "Genetic Stratigraphy and Geochronology of Last Interglacial Shorelines on the Central Coast of South Carolina." M.S. thesis, Louisiana State University, 2006.

Index